Nonlinear Dynamics
in Polymeric Systems

About the Cover

The cover images represent false-color infrared and visual images of "spin modes" of propagating fronts of free-radical polymerization of a multifunctional acrylate.

Courtesy of John A. Pojman

ACS SYMPOSIUM SERIES **869**

Nonlinear Dynamics in Polymeric Systems

John A. Pojman, Editor
University of Southern Mississippi

Qui Tran-Cong-Miyata, Editor
Kyoto Institute of Technology

**Sponsored by the
ACS Divisions of Polymer Chemistry, Inc. and
Physical Chemistry**

American Chemical Society, Washington, DC

Library of Congress Cataloging-in-Publication Data

Nonlinear dynamics in polymeric systems / John A. Pojman, editor, Qui Tran-Cong-Miyata, editor ; sponsored by the ACS Divisions of Polymer Chemistry, Inc. and Physical Chemistry.

    p. cm.—(ACS symposium series ; 869)

    Includes bibliographical references and index.

    ISBN 0–8412–3850–2

    1. Polymers—Congresses. 2. Polymerization–Congresses. 3. Molecular dynamics——Congresses. 4. Nonlinear theories—Congresses.

    I. Pojman, John A. (John Anthony), 1962- II. Tran-Cong-Miyata, Qui, 1952- III. American Chemical Society. Division of Polymer Chemistry, Inc. IV. American Chemical Society. Division of Physical Chemistry. V. American Chemical Society. Meeting (224th : 2002 : Boston, Mass.) VI. Series.

QD380.N66   2003
547′.7—dc22   2003060006

The paper used in this publication meets the minimum requirements of American National Standard for Information Sciences—Permanence of Paper for Printed Library Materials, ANSI Z39.48–1984.

PRINTED IN THE UNITED STATES OF AMERICA

Foreword

The ACS Symposium Series was first published in 1974 to provide a mechanism for publishing symposia quickly in book form. The purpose of the series is to publish timely, comprehensive books developed from ACS sponsored symposia based on current scientific research. Occasionally, books are developed from symposia sponsored by other organizations when the topic is of keen interest to the chemistry audience.

Before agreeing to publish a book, the proposed table of contents is reviewed for appropriate and comprehensive coverage and for interest to the audience. Some papers may be excluded to better focus the book; others may be added to provide comprehensiveness. When appropriate, overview or introductory chapters are added. Drafts of chapters are peer-reviewed prior to final acceptance or rejection, and manuscripts are prepared in camera-ready format.

As a rule, only original research papers and original review papers are included in the volumes. Verbatim reproductions of previously published papers are not accepted.

ACS Books Department

Contents

Frontal Polymerization

Interfacial Systems

Phase Separation

Oscillatory Systems

Indexes

Preface

In 1991, we editors met at a Gordon Research Conference on *Oscillations and Dynamic Instabilities in Chemical Systems*. Of the more than 120 participants, we were the only two who were working on polymeric systems. It was our dream that some day we would hold a symposium dedicated to nonlinear dynamics and polymers. This book is a realization of that dream.

This book is based on presentations made during the symposium *Nonlinear Dynamics in Polymeric Systems* held at the 224th National Meeting of the American Chemical Society (ACS) in Boston, Massachusetts on August 18–22, 2002, which was cosponsored by the ACS Divisions of Polymer Chemistry, Inc. and Physical Chemistry. More than 30 participants presented their work.

This volume brings together two fields of science that have, until recently, interacted only rarely. The central role of polymer science in understanding the substances of modern life and, indeed, of life itself is obvious. Nonlinear dynamics is a newer science, but one that has afforded important insights into phenomena in chemistry, physics, mathematics, biology, geology, and even the social sciences. Perhaps because so much of polymer science occurs in an industrial context, where nonlinear behavior such as chaos is typically seen as something to be avoided, even at the cost of suboptimal operating conditions, the amount of cross-fertilization between these areas has been limited. The purpose of this volume is to bring together practitioners in both fields to discuss the progress that has been made in unveiling and exploiting the nonlinear dynamical aspects of polymer systems and to identify problems that merit further study. A "Focus Issue" of the journal *Chaos* (*1*) and an issue of *Macromolecular Symposia* (*2*) provide an earlier look at this subject.

The two audiences for this book are nonlinear dynamicists who are interested in learning about polymers and polymer researchers who are interested in learning about nonlinear dynamics. We hope that researchers in each field will be able to appreciate the other field and to become inspired to begin their own research. Barriers currently exist for both groups. Few nonlinear dynamics researchers will have a background in polymers, especially if they come from a mathematical background and few polymer researchers will know of nonlinear dynamics, especially if they come from a chemistry back-ground. We hope this work will be a tool to bridge the gaps.

It is most exciting that most of the 'usual suspects' of nonlinear dynamics are present (i.e., temporal oscillations, chemical waves, propagating fronts, bifurcation analysis, spatial pattern formation, the Belousov–Zhabotinsky reaction, and interfacial instabilities. What distinguishes much of the work in this book from usual nonlinear dynamics is the goal of making some useful materials and devices. But even more distinguishes the works. New nonlinear phenomena arise in polymers, such as chemomechanical coupling in gels and phase separation induced by periodic forcing.

We are optimistic for the future of this field because so many interested works are present even though only a small fraction of polymer science is represented. We are confident that as more polymeric systems are explored with the tools of nonlinear dynamics, more exciting and usual phenomena will be discovered.

We hope the work inspires the reader to begin or continue in this new research area.

We thank the Iwatani Foundation for the financial support of the Symposium.

References

1.   *Chaos* Epstein, I. R; Pojman, J. A., editors, **1999**, *9*, 255–347.

2. *Nonlinear Dynamics in Polymer Science (PolyNon '99)*, Khokh-lov, A. R.; Tran-Cong-Miyata, Q.; Davydov, V. A.; Semion I. Kuchanov; Yamaguchi, T., Editors; Macromolecular Symposia; Wiley-VCH: Weinheim, Germany, 2000; volume16.

John A. Pojman
Department of Chemistry and Biochemistry
The University of Southern Mississippi
Hattiesburg, MS 39406–5043
(601) 266–5035 (telephone)
(425) 740–8514 (fax)
john@pojman.com

Qui Tran-Cong-Miyata
Department of Polymer Science and Engineering
Kyoto Institute of Technology
Matsugasaki Sakyo-ku Kyoto 606
Japan
81–75–724–7862 (telephone)
81–75–724–7710 (fax)
qui@ipc.kit.ac.jp

Nonlinear Dynamics
in Polymeric Systems

Background

Chapter 1

Nonlinear Dynamics and Polymeric Systems: An Overview

Irving R. Epstein[1], John A. Pojman[2], and Qui Tran-Cong-Miyata[3]

[1]Department of Chemistry and Volen Center for Complex Systems, MS 015, Brandeis University, Waltham, MA 02454
[2]Department of Chemistry and Biochemistry, University of Southern Mississippi, Hattiesburg, MS 39406
[3]Department of Polymer Science and Engineering, Kyoto Institute of Technology, Matsugasaki, Sakyo-ku, Kyoto 606–8585, Japan

This chapter provides an overview of nonlinear chemical dynamics and synthetic polymeric systems, the progress that has been made in bringing these areas together, and some issues that merit further study. Some key themes and results of nonlinear chemical dynamics are cited, and an attempt is made to link them to important questions in polymeric systems.

Nonlinear Chemical Dynamics

Nonlinear dynamics is a vast field, originating in mathematics and physics. we shall focus here on its chemical aspects. In this section, we provide a brief overview of the key concepts and phenomena relevant to chemical systems. The next section considers nonlinear dynamics in polymeric systems. Finally, we consider future directions for the field. A more detailed treatment of nonlinear chemical dynamics, including a more detailed discussion of polymer systems, may be found in reference (1).

2

Some key notions in that arise in nonlinear chemical dynamics may be briefly summarized as follows:

a)   Behavior of interest occurs *far from equilibrium*. While systems near equilibrium are more familiar and frequently more tractable, since they can often be approximated as linear, they generate a much scantier set of behaviors. Often, the system is maintained far from equilibrium by keeping it *open*, i.e., by maintaining flows of material and/or energy into and out of the system.

b)   The behaviors of interest are *self-organized*, in that they arise spontaneously without requiring a template or a seed.

c)   *Feedback* typically plays a key role, often in the form of *autocatalysis*, in which the product of a reaction influences the rate of that reaction.

d)   Many of the systems of interest display *multiple stable states*, i.e., the system can exist in two or more dynamically different states (steady state, oscillation, chaos) under the same set of externally imposed conditions (e.g., input concentrations, temperature). The state that is actually obtained depends upon the history of the system – there is a memory effect or *hysteresis*.

e)   A small number of key *control* or *bifurcation parameters* determine the state of the system. These are typically input concentrations of key reactants, the temperature of the system, or the rate at which material flows through the system. As one of these parameters is varied, an *instability* may occur, so that a previously stable state loses its stability and the system undergoes a transition to a different, stable state.

f)   Nonlinear chemical systems often exhibit *spatio-temporal structure* in the form of periodic oscillation of concentrations in time and/or in space or traveling waves of concentration.

g)   Certain behaviors display *universality*, i.e., qualitatively similar phenomena, such as spiral waves or period doubling as a system approaches chaos, occur in systems that would appear to be quite different mechanistically.

h)   Many nonlinear systems consisting of multiple components exhibit the phenomenon of *emergence*, whereby the behavior of the whole system is considerably more complex than, and difficult or impossible to deduce from, the behaviors of the individual components.

We list below some of the most interesting and most thoroughly studied phenomena that arise in nonlinear chemical dynamics.

4

Chemical Oscillations

The study of nonlinear chemical dynamics begins with chemical oscillators – systems in which the concentrations of one or more species increase and decrease periodically, or nearly periodically. While descriptions of chemical oscillators can be found at least as far back as the nineteenth century (and chemical oscillation is, of course, ubiquitous in living systems), systematic study of chemical periodicity begins with two accidentally discovered systems associated with the names of Bray (2) and of Belousov and Zhabotinsky (BZ) (3,4). These initial discoveries were met with skepticism by chemists who believed that such behavior would violate the Second Law of Thermodynamics, but the development of a general theory of nonequilibrium thermodynamics (5) and of a detailed mechanism (6) for the BZ reaction brought credibility to the field by the mid-1970's. Oscillations in the prototypical BZ reaction are shown in Figure 1.

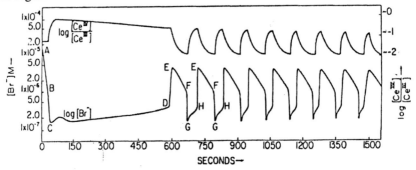

Figure 1. Oscillatory behavior following an initial induction period in the potentials of a platinum redox electrode and a bromide-sensitive electrode in a BZ reaction mixture containing an aqueous solution of bromate, malonic acid, a ceric salt, and sulfuric acid. Reproduced from Field et al. (6). Copyright 1972 American Chemical Society.

The development of a systematic approach to designing chemical oscillators (7) has resulted in the discovery of several dozen new oscillating reactions. Mechanisms have been developed for many, and their simulated kinetics shows good agreement with the experimentally observed behavior. There is even a "taxonomy" of chemical oscillators, in which reactions with similar chemistry are classified into families. The work described in this section mainly involves "well-stirred" systems, which are assumed to be homogenous, but more detailed studies (8) reveal a significant dependence of oscillatory behavior on the (always imperfect) quality of mixing. Another aspect of spatial inhomogeneity is described in the next section.

Waves and Patterns

One of the most striking features of the prototype chemical oscillator, the BZ reaction, is that an apparently uniform solution unstirred in a thin layer spontaneously generates a striking pattern consisting of sets of concentric rings ("target patterns") or rotating spirals (9). An example is shown in Figure 2. Similar patterns are seen in other chemical oscillators as well as in a variety of biological systems, including aggregating slime molds and developing frog oocytes. The behavior of these and other traveling "trigger" waves in reaction-diffusion systems can be understood in terms of the interaction between species that react on quite different time scales (10).

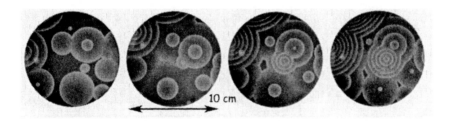

Figure 2. Target patterns in the BZ reaction. Four successive snapshots taken after reaction evolves from initially homogeneous dark solution.

Another set of pattern formation phenomena involve stationary, or Turing patterns (11), which arise in systems where an inhibitor species diffuses much more rapidly than an activator species. These patterns, which are often invoked as a mechanism for biological pattern formation, were first found experimentally in the chlorite-iodide-malonic acid reaction (12). Examples of typical spot and stripe patterns appear in Figure 3. Recently, experiments in reverse microemulsions have given rise not only to the waves and patterns described above, but to a variety of novel behaviors, including standing waves and inwardly moving spirals, as well (13).

More Complex Phenomena

"Simple" periodic chemical oscillation may now be said to be reasonably well understood. More complex behavior can arise when a single oscillator is pushed into new realms or when it is coupled either to other oscillators or to external influences. Some chemical oscillators that are simply periodic under one set of conditions can exhibit complex, multi-peaked, periodic or even aperiodic, chaotic (14) behavior at other concentrations and flow rates in an open reactor. Some examples of chaotic oscillations in the chlorite-thiosulfate system are shown in Figure 4. Coupling two or more reactions together can result in the

6

generation of oscillations in a previously quiescent system, the cessation of oscillations in a pair of formerly oscillating systems, or in a range of complex periodic and aperiodic oscillatory modes. Chemical oscillators containing charged or paramagnetic species are affected by electric (15,16) or magnetic (17) fields and even, in the case of traveling waves, by gravitational fields (18), as the chemical changes become coupled to convective motion.

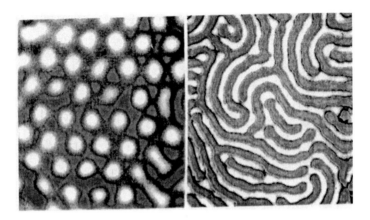

Figure 3. Turing patterns in the chlorite-iodide-malonic acid reaction. Dark areas show high concentrations of starch-triiodide complex. Each frame is approximately 1.3 mm square. Images courtesy of Patrick De Kepper.

Polymeric Systems

What is to be gained from applying the methods and concepts of nonlinear dynamics to polymer systems? Are there things that nonlinear dynamicists can learn, or that polymer scientists can make, that would not be possible without bringing these two apparently disparate fields into contact? First we briefly review some distinguishing characteristics of polymers. Next, we suggest three challenges that present themselves. We then examine sources of feedback in polymeric systems. Next, we propose several approaches to develop nonlinear dynamics with polymers. Finally, we give some examples of results that suggest that these approaches are likely to bear fruit.

We do not have the space to review polymers but refer the reader to several texts (19-21). What we seek to provide here is a brief overview of the most important differences between polymeric systems and small molecule ones, review sources of feedback, approaches to nonlinear dynamics with polymers, including some specific examples.

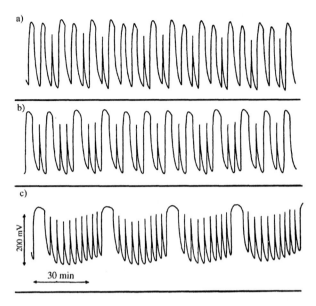

Figure 4. Chaotic oscillation in the potential of a Pt redox electrode in the chlorite-thiosulfate reaction in a flow reactor. Note the aperiodic alternation between large and small amplitude peaks. Input concentrations, $[ClO_2^-]_0 = 5 \times 10^{-4}$ M, $[S_2O_3^{2-}]_0 = 3 \times 10^{-4}$ M, pH = 4, residence time in reactor = a) 6.8 min, b) 10.5 min, c) 23.6 min. Reproduced from Orbán et al. (14). Copyright 1982 American Chemical Society.

We will not deal with biological systems, which certainly are polymeric systems. Biological systems are nonlinear dynamical systems but they are complex enough and so important that they warrant separate treatment. We refer the reader to reference (22) for an introduction to the topic.

What is Special About Polymers?

We briefly discuss the differences between polymers and the typical inorganic systems most often studied in nonlinear chemical dynamics. The distinguishing feature of polymers is their high molecular weight. The simplest synthetic polymer consists from hundreds to even millions of a single unit, the monomer, that is connected end to end in a linear chain. However, a distribution of chain lengths always exists in a synthetic system. The molecular weight distribution can be quite broad, often spanning several orders of magnitude of molecular weight.

Linear polymers, are often *thermoplastic*, meaning they can flow at some temperature, which depends on the molecular weight, e.g., polystyrene. Polymers need not be simple chains but can be branched or networked. Crosslinked polymers can be gels that swell in a solvent or *thermosets,* which form rigid 3-dimensional networks when the monomers react, e.g., epoxy resins. This interconnectedness allows long-range coupling in the medium.

The physical properties of the reaction medium change dramatically during reaction. For example, the viscosity almost always increases orders of magnitude. These changes often will affect the kinetic parameters of the reaction and the transport coefficients of the medium.

Phase separation is ubiquitous with polymers. Miscibility between polymers is the exception.

Challenges

In contemplating the possible payoffs from applying nonlinear dynamics to polymeric systems, one might ask

a)   Are there new materials that can be made by deliberately exploiting the far-from-equilibrium behavior of processes in which polymers are generated?

b)   Are there existing materials and/or processes that can be improved by applying the principles and methods of nonlinear dynamics?

c)   Are there new nonlinear dynamical phenomena that arise because of the special properties of polymer systems?

Sources of Feedback

Synthetic polymer systems can exhibit feedback through several mechanisms. The simplest is thermal autocatalysis, which occurs in any exothermic reaction. The reaction raises the temperature of the system, which increases the rate of reaction through the Arrhenius dependence of the rate constants. In a spatially distributed system, this mechanism allows propagation of thermal fronts. Free-radical polymerizations are highly exothermic .

Free-radical polymerizations of certain monomers exhibit autoacceleration at high conversion via an additional mechanism, the isothermal "gel effect" or "Trommsdorff effect" (*23-26*). These reactions occur by the creation of a radical that attacks an unsaturated monomer, converting it to a radical, which can add to another monomer, propagating the chain. The chain growth terminates when two radical chains ends encounter each other, forming a stable chemical bond. As the polymerization proceeds, the viscosity increases. The diffusion-limited termination reactions are thereby slowed down, leading to an increase in the overall polymerization rate. The increase in the polymerization rate induced by the increase in viscosity builds a positive feedback loop into the polymerizing system.

Some polymer hydrogels exhibit "phase transitions" as the pH and/or temperature are varied (27,28). The gel can swell significantly as the conditions are changed and can also exhibit hysteresis (28,29).

Most polymers are not miscible. Introducing chemical reactions to an initially miscible polymer mixture often leads to phase separation (30). Autocatalytic behavior driven by tchemical reactions and concentration fluctuations in miscible polymer mixtures was recently found in photo-cross-linked polymer mixtures (31). Concentration fluctuations increase as the reaction proceeds, leading to the condensation of photoreactive groups labeled on one of the polymer components. This condensation leads to an increase in the reaction yield that, in turn, accelerates the concentration fluctuations. . A positive feedback can thus be built in the reacting mixture under appropriate conditions.

If two immiscible polymers are dissolved in a common solvent, which is then removed by evaporation, phase separation will occur. If the solvent is removed rapidly, non-equilibrium patterns may result (32).

The necking phenomenon observed upon stretching a polymer film at a constant temperature is a well-known consequence of a negative feedback loop driven by the interplay between the increase in temperature associated with the sample deformation and its glassification caused by the heat exchange with the environment (33). Oscillatory behavior and period-doubling in the stress resulting from a constant strain rate have been experimentally observed.

Diffusion of small molecules, usually solvents, into glassy polymers exhibits 'anomalous' or 'non-Fickian' behavior (34). As the solvent penetrates, the diffusion coefficient increases because the glass transition temperature is lowered. The solvent acts as a plasticizer, increasing the free volume and the mobility of the solvent. Thus we have an autocatalytic diffusion process. This can be relevant in Isothermal Frontal Polymerization, which we discuss below.

Finally, polymer melts and solutions are usually non-Newtonian fluids (35-37). They often exhibit shear thinning, which means the viscosity decreases as the shear is increased. This can lead to unusual phenomena. For example, when a polymer melt is extruded through a die, transient oscillations can occur (38,39). (Polymers can also exhibit shear thickening)

An unusual phenomenon is the Weissenberg effect, or the climbing of polymeric liquids up rotating shafts (37). A Newtonian fluid, on the other hand, is depressed by rotation because of centrifugal forces.

Approaches to Nonlinear Dynamics in Polymeric Systems

We propose three approaches to creating nonlinear dynamical systems with polymers:

a)   Couple polymers and polymer-forming reactions to other nonlinear systems (Type I)

b)   Create a dynamical system using the inherent nonlinearities in polymeric systems (Type II)

c)   Polymer systems are invariably characterized by polydispersity of the molecular weight distribution. One should be able to exploit the distribution of polymer lengths to amplify nonlinear effects in polymer systems, perhaps because of the molecular weight dependence of the diffusion coefficient. We know of no experimental work but there has been a theoretical work considering such an effect on ester interchange reactions (40).

Type I: Coupling to another nonlinear system

Given the importance of the BZ reaction in nonlinear chemical dynamics, it is not surprising that polymers and polymerizations would be coupled to it. Pojman et al. studied the BZ reaction to which acrylonitrile was added and showed that the polyacrylonitrile was produced periodically in phase with the oscillations (41). Given that radicals are produced periodically from the oxidation of malonic acid by ceric ion, it seemed reasonable to assume the periodic appearance of polymer was caused by periodic initiation. However, Washington et al. showed that periodic termination by bromine dioxide caused the periodic polymerization (42).

This system is not in itself useful because the polymer is brominated and the reagents expensive but such a system has been used to generate periodic self-assembly of aggregates of acrylonitrile-derivatized gold nanocrystals (43). Kalyshyn et al. reported macroscopic patterns in crosslinked polyacrylamide with BZ waves (44).

Pelle et al. have examined how water-soluble alcohols affect the BZ system (45). Lombardo et al. studied the effects of polyethylene glycol on the BZ system (chapter 21 of this volume).

Li et al. used a polymerization reaction to 'fix' patterns generated by convection (46).

An exciting application of coupling to another nonlinear system was demonstrated by Yoshida et al. who created a self-oscillating gel by coupling a pH oscillating reaction with a polymeric gel that expands and contracts with changes in pH (*47*). They have also used a gel in which the ruthenium catalyst of the BZ reaction is chemically incorporated into polymer (*48*).

There is not always a clear distinction between Type I and Type II systems. Section II of this volume addresses several works with gels. In some cases the nonlinearities of the gel also play a role. The extreme case is the system developed by Siegel. He and his colleagues utilized the hysteresis in a hydrogel's permeability to create autonomous chemomechanical oscillations in a hydrogel/enzyme system driven by glucose (*49,50*). This is also addressed in chapter 4.

Type II: Using the Inherent Nonlinearities in a Polymer System

Oscillations in a CSTR

With their combination of complex kinetics and thermal, convective and viscosity effects, polymerizing systems would seem to be fertile ground for generating oscillatory behavior. Despite the desire of most operators of industrial plants to avoid nonstationary behavior, this is indeed the case. Oscillations in temperature and extent of conversion have been reported in industrial-scale copolymerization (*51*).

Teymour and Ray reported both laboratory-scale CSTR experiments (*52*) and modeling studies (*53*) on vinyl acetate polymerization. The period of oscillation was long, about 200 minutes, which is typical for polymerization in a CSTR. Papavasiliou and Teymour review nonlinear dynamics in CSTR polymerizations in chapter 22.

Emulsion polymerization as well has been found to produce oscillations in both the extent of conversion and the surface tension of the aqueous phase (*54*).

Frontal Polymerization

One of the most promising applications of nonlinear dynamics to polymer science is the phenomenon of frontal polymerization (Section III). Frontal polymerization is a process of converting monomer into polymer via a localized reaction zone that propagates through the monomer. There are two modes of frontal polymerization.

Thermal frontal polymerization involves the coupling of thermal diffusion and Arrhenius reaction kinetics of an exothermic polymerization (*55*). Thermal frontal polymerization has promise for making specialized materials in which the rapid reaction is valuable (*56-58*) or for which a special gradient is needed

(*59-61*). Curing of thick composites may also be a useful application (*62,63*). In chapter 9, Pojman et al. review nonlinear dynamics in frontal polymerization.

Isothermal Frontal Polymerization (IFP), also called Interfacial Gel Polymerization, is a slow process in which polymerization occurs at a constant temperature and a localized reaction zone propagates because of the gel effect (*64,65*). Using IFP (*27*), one can control the gradient of an added material like a dye, to generate materials useful, for example, in optical applications (*66,67*). Lewis et al. provide experimental and theoretical results in chapter 14.

Dewetting Phenonema

When a polymer solution is allowed to evaporate from a surface, dewetting instabilities can arise, producing modulated patterns like spinodal structures (*68*). Section IV addresses the applicability of these interesting phenomena to making opto-electronical devices.

Spatial Pattern Formation

Typical phase separation leads to a two-phase disordered morphology. Multiphase polymeric materials with a variety of co-continuous structures can be prepared by controlling the kinetics of phase separation via spinodal decomposition using appropriate chemical reactions. By taking advantages of photo-crosslinking and photoisomerization of one polymer component in a binary miscible blend, Tran-Cong-Miyata and coworkers (*69*) have been able to prepare materials, known as semi-interpenetrating polymer networks, and polymers with co-continuous structures in the micrometer range. These structures are similar to the so-called Turing structure. Section V provides more information on phase separation.

Conclusions

The application of nonlinear dynamics to polymer systems has enormous untapped potential for the production of novel materials with desirable structures and properties. Exploiting these opportunities will require that polymer scientists become better acquainted with the phenomena and techniques of nonlinear dynamics and that practitioners of nonlinear dynamics familiarize themselves with the properties and capabilities of polymeric systems. The topics sketched here provide only a hint of the myriad future possibilities. The chapters that comprise this book add more flesh to these bare bones, and it is already apparent that there are many challenging and exciting problems to be solved.

Acknowledgements

IRE acknowledges support from the Chemistry Division of the National Science Foundation. and thank his colleagues, Milos Dolnik, Vladimir Vanag, Lingfa Yang and Anatol Zhabotinsky, for their many contributions. JAP acknowledges support from We acknowledge support from the NASA Reduced Gravity Materials Science Program (NAG 8-973 and NAG 8-1858) and the National Science Foundation (CTS-0138660). QTCM would like to appreciate the financial support from MONKASHO (Ministry of Education, Japan) through the Grant-in-Aid No. 13031054 on the Priority-Area-Research *"Dynamic Control of Strongly Correlated Soft Materials"*.

References

1. Epstein, I. R.; Pojman, J. A. *An Introduction to Nonlinear Chemical Dynamics: Oscillations, Waves, Patterns and Chaos*; Oxford University Press: New York, 1998.
2. Bray, W. C. *J. Am. Chem. Soc.* **1921**, *43,* 1262-1267.
3. Belousov, B. P. *Sbornik Referatov po Radiatsionni Meditsine* **1958**, 145.
4. Belousov, B. P. A Periodic Reaction and Its Mechanism: in *Oscillations and Traveling Waves in Chemical Systems*; Field, R. J. Burger, M., Ed.; Wiley: New York, 1985; pp 605-613.
5. Glansdorff, P.; Prigogine, I. *Thermodynamics of Structure, Stability and Fluctuations*; Wiley: New York, 1971.
6. Field, R. J.; Körös, E.; Noyes, R. *J. Am. Chem. Soc.* **1972**, *94,* 8649-8664.
7. Epstein, I. R.; Kustin, K.; De Kepper, P.; Orbán, M. *Scientific American* **1983**, *248,* 112-123.
8. Roux, J. C.; De Kepper, P.; Boissonade, J. *Phys. Lett.* **1983**, *97A,* 168-170.
9. Zaikin, A. N.; Zhabotinskii, A. M. *Nature* **1970**, *225,* 535-537.
10. Fife, P. Propagator-Controller Systems and Chemical Patterns: in *Nonequilibrium Dynamics in Chemical Systems*; Vidal, C. Pacault, A., Ed.; Springer-Verlag: Berlin, 1984; pp 76-88.
11. Turing, A. M. *Philos. Trans. R. Soc. London Ser. B* **1952**, *237,* 37-72.
12. Castets, V.; Dulos, E.; Boissonade, J.; De Kepper, P. *Phys. Rev. Lett.* **1990**, *64,* 2953-2956.
13. Vanag, V. K.; Epstein, I. R. *Phys. Rev. Lett.* **2001**, *87,* 228-301.
14. Orbán, M.; Epstein, I. R. *J. Phys. Chem.* **1982**, *86,* 3907-3910.
15. Sevcikova, H.; Marek, M. *Physica D* **1983**, *9,* 140-156.

14

16. Sevcikova, H.; Marek, M. *Physica D* **1986**, *21*, 61-77.
17. Boga, E.; Kádár; Peintler, G.; Nagypál, I. *Nature* **1990**, *347*, 749-751.
18. Nagypál, I.; Bazsa, G.; Epstein, I. R. *J. Am. Chem. Soc.* **1986**, *108*, 3635-3640.
19. Odian, G. *Principles of Polymerization*; Wiley: New York, 1991.
20. Allcock, H. R.; Lampe, F. W. *Contemporary Polymer Chemistry*; Prentice-Hall: Englewood Cliffs, 1981.
21. Sperling, L. H. *Introduction to Physical Polymer Science*; John Wiley & Sons, INC.: New York, 1992.
22. Goldbeter, A. *Biochemical Oscillations and Cellular Rhythms: The Molecular Bases of Periodic and Chaotic Behaviour*; Cambridge University Press: Cambridge, UK, 1996.
23. Norrish, R. G. W.; Smith, R. R. *Nature* **1942**, *150*, 336-337.
24. Trommsdorff, E.; Köhle, H.; Lagally, P. *Makromol. Chem.* **1948**, *1*, 169-198.
25. O'Neal, G. O.; Wusnidel, M. B.; Torkelson, J. M. *Macromolecules* **1998**, *31*, 4537-4545.
26. O'Neal, G. O.; Wusnidel, M. B.; Torkelson, J. M. *Macromolecules* **1996**, *29*, 7477-7490.
27. Tanaka, T. *Phys. Rev. Lett.* **1978**, *40*, 820-823.
28. Addad, J. P. C. *Physical Properties of Polymer Gels*; Wiley: Chichester, 1996.
29. Hirotsu, S.; Hirokawa, Y.; Tanaka, T. *J. Chem. Phys.* **1987**, *87*, 1392.
30. Tran-Cong, Q.; Harada, A. *Phys. Rev. Lett.* **1996**, *76*, 1162-1165.
31. Tran-Cong, Q.; Harada, A.; Kataoka, K.; Ohta, T.; Urakawa, O. *Phys. Rev. E* **1997**, *55*, R6340-R6343.
32. Edel, V. *Macromolecules* **1995**, *28*, 6219-6228.
33. Andrianova, G. P.; Kechkyan, A. S.; Kargin, V. A. *J. Poly. Sci.* **1971**, *9*, 1919-1933.
34. Crank, J. *Mathematics of Diffusion*; Clarendon: Oxford, 1975.
35. Severs, E. T. *Rheology of Polymers*; Reinhold: New York, 1962.
36. Ferry, J. D. *Viscoelastic Properties of Polymers*; Wiley: New York, 1980.
37. Gupta, R. K. *Polymer and Composite Rheology*; Marcel Dekker: New York, 2000.
38. Meissner, J. *J. Appl. Polym. Sci.* **1972**, *16*, 2877-2899.
39. Bird, R. B.; Armstrong, R. C.; Hassager, O. *Dynamics of Polymeric Liquids, 2nd ed. Vol. 1*; Wiley: New York, 1987.
40. Pojman, J. A.; Garcia, A. L.; Kondepudi, D. K.; Van den Broeck, C. *J. Phys. Chem.* **1991**, *95*, 5655-5660.
41. Pojman, J. A.; Leard, D. C.; West, W. *J. Am. Chem. Soc.* **1992**, *114*, 8298-8299.
42. Washington, R. P.; Misra, G. P.; West, W. W.; Pojman, J. A. *J. Am. Chem. Soc.* **1999**, *121*, 7373-7380.

15

43. Dylla, R. J.; Korgel, B. A. *ChemPhysChem* **2001**, *1*, 62-64.
44. Kalishyn, Y. Y.; Khavrus, V. O.; Strizhak, P. E.; Seipel, M.; Münster, A. F. *Chem. Phys. Lett.* **2002**, *363*, 534-539.
45. Pelle, K. W., M.; Noszticzius, Z.; Lombardo, R.; Sbriziolo, C.; Turco Liveri, M.L. *J. Phys. Chem. A* **2003**, 2039-47.
46. Li, M.; Xu, S.; Kumacheva, E. *Macromolecules* **2000**, *33*, 4972-4978.
47. Yoshida, R.; Ichijo, H.; Hakuta, T.; Yamaguchi, T. *Macromol. Rapd Commun.* **1995**, *16*, 305-310.
48. Yoshida, R.; Onodera, S.; Yamaguchi, T.; Kokufuta, E. *J. Phys. Chem. A* **1999**, *103*, 8573-8578.
49. Dhanarajan, A. P.; Misra, G. P.; Siegel, R. A. *J. Phys. Chem. A.* **2002**, *106*, 8835-8838.
50. Leroux, J.-C.; Siegel, R. A. *Chaos* **1999**, *9*, 267-275.
51. Keane, T. R. Single-Phase Polymerization Reactors, *Chemical Reaction Engineering: Proceedings of the Fifth European/Second International Symposium on Chemical Reaction Engineering*, **1972**, A7-1 - A7-9.
52. Teymour, F.; Ray, W. H. *Chem. Eng. Sci.* **1992**, *47*, 4121-4132.
53. Teymour, F.; Ray, W. H. *Chem. Eng. Sci.* **1992**, *47*, 4133-4140.
54. Schork, F. J.; Ray, W. H. *J. Appl. Poly. Sci. (Chem.)* **1987**, *34*, 1259-1276.
55. Pojman, J. A.; Ilyashenko, V. M.; Khan, A. M. *J. Chem. Soc. Faraday Trans.* **1996**, *92*, 2825-2837.
56. Khan, A. M.; Pojman, J. A. *Trends Polym. Sci. (Cambridge, U.K.)* **1996**, *4*, 253-257.
57. Nagy, I. P.; Sike, L.; Pojman, J. A. *J. Am. Chem. Soc.* **1995**, *117*, 3611-3612.
58. Gill, N.; Pojman, J. A.; Willis, J.; Whitehead, J. B. *J. Poly. Sci. Part A. Polym. Chem.* **2003**, *41*, 204-212.
59. Pojman, J. A.; McCardle, T. W. U.S. Patent 6,057,406,2000
60. Pojman, J. A.; McCardle, T. W. U.S. Patent 6,313,237,2001
61. Chekanov, Y. A.; Pojman, J. A. *J. Appl. Polym. Sci.* **2000**, *78*, 2398-2404.
62. Kim, C.; Teng, H.; Tucker, C. L.; White, S. R. *J. Comp. Mater.* **1995**, *29*, 1222-1253.
63. Chekanov, Y.; Arrington, D.; Brust, G.; Pojman, J. A. *J. Appl. Polym. Sci.* **1997**, *66*, 1209-1216.
64. Koike, Y.; Takezawa, Y.; Ohtsuka, Y. *Appl. Opt.* **1988**, *27*, 486-491.
65. Smirnov, B. R. M. k., S. S.; Lusinov, I. A.; Sidorenko, A. A.;Stegno, E. V.; Ivanov, V. V. *Vysokomol. Soedin., Ser. B* **1993**, *35*, 161-162.
66. Ishigure, T.; Nihei, E.; Koike, Y. *Appl. Opt.* **1994**, *33*, 4261-4266.
67. Masere, J.; Lewis, L. L.; Pojman, J. A. *J. Appl. Polym. Sci.* **2001**, *80*, 686-691.
68. Mitlin, V. S. *J. Colloid Interface Sci.* **1993**, *156*, 491-497.
69. Harada, A.; Tran-Cong, Q. *Macromolecules* **1997**, *30*, 1643-1650.

Chapter 2

Self-Organization of Hierarchy: Dissipative-Structure Assisted Self-Assembly of Metal Nanoparticles in Polymer Matrices

Tomohiko Yamaguchi[1,2,*], Nobuhiko Suematsu[1,2], and Hitoshi Mahara[1]

[1]Nanotechnology Research Institute, National Institute of Advanced Industrial Science and Technology, AIST Central 5-2, 1-1-1 Higashi, Tsukuba 305-8565, Japan
[2]Graduate School of Pure and Applied Sciences, University of Tsukuba, 1-1-1 Tennoudai, Tsukuba 305-8577, Japan

Though thermodynamically different, self-assembly and dissipative structure formation often work together to bring about highly ordered structures in an open system. This mutual assistance between self-assembly and dissipative structure formation is regarded as self-organization for a system to increase the degrees of hierarchy and complexity. An example for this comprehensive idea is given by hierarchic self-organization of organo-passivated metal nanoparticles in dissipatively isolated polymer matrices.

Introduction

The concept of self-organization is originated by Schrödinger *(1)*, who reconstructed the long issue on life as a problem of order in a living system. He thought that an order in a living system was brought about via *order from order* and *order out of disorder*. Order from order is related to the structure of biomolecules such as DNA and to the openness of the system as well (i.e., getting order from the surroundings as *negative entropy* to maintain its order). Order out of disorder is related to spatio-temporal pattern formation under the conditions far from equilibrium. The latter is named as *dissipative structure* by Prigogine *(2)* and has been studied intensively since 1970s. On the other hand, there has been extensive studies on molecular biology and on biomimetic chemistry such as host-guest chemistry and supramolecular chemistry, and self-organization in chemistry is often referred to as *self-assembly* of (tailor-made) molecules in order to realize a higher hierarchy via non-covalent intermolecular forces between molecular components *(3)*.

Self-Organization in Chemistry

Self-Assembly and Dissipative Structure

Thus, we know two principles of self-organization: self-assembly near equilibrium conditions and dissipative structure formation under conditions far from equilibrium. As summarized in Table 1, these are considerably different in time, scale, order of driving force, existence of potential function, and so on. Because of these thermodynamical differences and their historical backgrounds, they seem to have been studied almost independently in different research fields.

Table 1. Self-Assembly and Dissipative Structure

	Self-Assembly	*Dissipative Structure*
Periodicity	Spatial	Spatial and temporal
Wavelength[*]	10^0-10^1	10^2-10^6
Driving force[**]	10^1	10^2
Entropy production	Minimum	No universality
Potential function	Exists	Not known
Reversible	Yes	No
Described by	Phase transition	Instability and bifurcation

[*] With respect to the size of components.
[**] With respect to thermal noise.

Advantages of Each Self-Organization

A self-assembled structure is a static and stable structure against thermal fluctuations. Recent progress in molecular chemistry enables us to design a self-assembled supramolecular structure by designing and synthesizing component molecules, and handling them like pieces of toy blocks. Driving forces of self-assembly are hydrogen bondings, salt bridges and solvation forces. Geometrical and topological characteristics of molecules enhance these forces. Self-assembly requires no introduction of energy or matter into the system of concern under thermal equilibrium conditions. In most practical cases, however, self-assembly takes place near equilibrium conditions, where continuous flows of heat and chemical reaction exist. These thermodynamical fluxes must remain at linear regime because of the constraint of minimum entropy production. Theoretically, the most stable structure should be selected; however, it is still very difficult to predict even crystallographic structures from given structures of organic compounds (4).

Dissipative structures have some interesting characteristics from the viewpoint of pattern formation:

1. The global structure of dissipative structure may be controlled by its local structure. It is because of entrainment between coupled oscillators distributed in space. As shown in Figure 1, for example, the shape of a spiral in the Belousov-Zhabotinsky (BZ) reaction is reversibly transformed from Archimedean to logarithmic by changing the core size of the spiral (5). It is a kind of information processing in dissipative structure (transduction and amplification of external information). Mathematically, the core includes the singular point where the phase of the periodic reaction is not defined. This nature of global control through the singular point is universal among so-called excitable media with temporal periodicity.

2. Stationary patterns are not specific to self-assembly. Turing structures (6) are time-independent dissipative structures. Their characteristic wavelengths are intrinsically determined by the reaction rates and the diffusion coefficients, and are independent of boundary conditions. In addition to an ordinary route to the Turing structure via global noise-driven diffusion-induced instability, it has been found recently a unique route through the self-duplication of spots that finally cover the whole active medium to result in a stationary Turing structure (Figure 2) (7).

3. Some of spatio-temporal patterns and time-evolving patterns in dissipative structures can be frozen to obtain stationary structure. Addition of Ag^+ ions in the BZ reaction records the spiral pattern (8), and quenching the photoinduced phase separation in a polymer mixture results in a Turing-type structure (9).

Self-Organization of Hierarchy by Mutual Assistance between Self-Assembly and Dissipative Structure

Under many practical situations such as in biological systems, above-mentioned two self-organization principles are not always distinguishable so clearly. Contrary, they often play their roles in different spatio-temporal scales to bring about highly complicated structures in open systems. A biological system is a good but not the only example; we can find it in a very simple system as well.

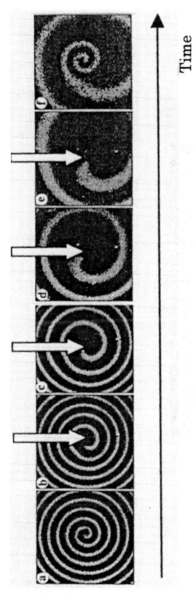

Figure 1. Control of the spiral shape in the oscillatory Belousov-Zhabotinsky reaction. Ru(bpy)$_3^{2+/3+}$ was used as catalyst. Ar$^+$ laser beam was irradiated (illustrated by white arrow) at the core of the rotating spiral to increase the size of the core region. The morphology of spiral changed reversibly from Archimedean to logarithmic, and the wave profile from trigger-wave to phase-wave. Controlling global structure by local control of singular region is characteristic in dissipative structures.

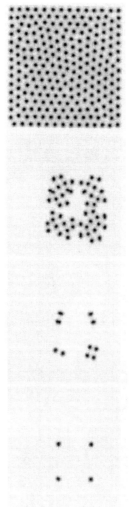

Figure 2. A new route to reach to a stationary structure (the Turing structure) via self-duplication of dots. Gray-Scott model (10) is used for calculations. The last panel is still not symmetrical because of the influence of noise initially added randomly at each pixel.

For example, one can see dissipative structure-assisted self-assembly of molecules or ions in dendritic crystal growth. When the driving force (the degree of supercooling) is large enough and the system is set far from the equilibrium (i.e., the melting point), the smoothly growing surface of a crystal becomes no more stable and is replaced by a new mode of dendritic growthing surface. Its characteristic wavelength is subject to the transport of latent heat from the growing interface. This is known as Mullins-Sekerka instability *(11)*. The interface of a growing dendritic crystal provides an open system to support the dissipative structure with its characteristic wavelength(s), and the molecules or ions incorporated into the growing crystal are regarded to be self-assembled into the specific coordinate of dendritic crystal by the assistance of the dissipative structure.

This complex process of dissipative structure-assisted self-assembly plays an important role for pattern formation in biological systems and many other artificial systems. So far as the authors know, this concept was firstly proposed by Lefever in 2000 *(12)*. We may extend his idea to think of an alternative concept: a self-assembly-assisted dissipative structure. It is obvious that most biological dissipative structures such as traveling waves along giant axons and

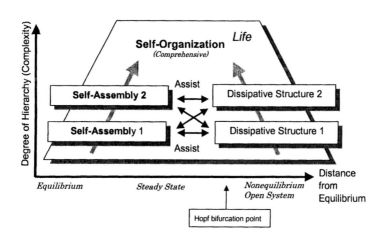

Figure 3. Two well-known principles of self-organization, i.e., self-assembly and dissipative structure, and a new concept of comprehensive self-organization. The axis of abscissa indicates the distance from the thermal equilibrium. Temporal dissipative structure appears at the Hopf bifurcation point. Self-assembly and dissipative structure are complimentary, and their mutual assistance makes it easier for the system to step-up the stair of hierarchy and complexity. Both dissipative structure-assisted self-assembly and self-assembly-assisted dissipative structure are possible.

wavy beating of cilia are realized on the basis of self-assembled substructure of biomolecules.

What is the fruit of this complicated picture of self-organization? The answer is the increase in degree of hierarchy and complexity (see Figure 3). As shown in Table 1, self-assembly and dissipative structure are in a complementary relationship in structural, temporal, and conditional senses. This complementarity is the basis of the mutual assistance between self-assembly and dissipative structure formation. We might increase the complexity of a system by self-assembly scenario only or by the scenario of dissipative structure formation only. We might increase the number of hierarchic levels by these independent scenarios as well. But in some cases (or in most cases) the comprehensive scenario of self-organization based on the mutual assisting relation will provide more simple and powerful strategies for tangling up the stair of hierarchy. It could be a strategy to be chosen by abiotic systems towards the origins of life (the prebiotic chemical evolution), and succeeding evolution of life. This strategy would answer the request of bottom-up procedures in recent nanotechnology. Therefore, recent interest among researchers in materials science has been focused on the mutual assisting relations between self-assembly and dissipative structure *(13)*. Figure 3 summarizes the present idea, and below we will give our original results as an example relevant to materials science.

Self-Organization of Hierarchic Structure

Metal Nanoparticle As Conducting Material

In 1981-1986, a pioneering research on metal nanoparticles had been carried out in Tsukuba, Japan. This project was called Hayashi Ultrafine Particle Project. The researchers interest was not only directed to pure science but practical application of metal nanoparticles. They succeeded in producing well-defined crystalline metal nanoparticles by high-temperature evaporation of Au, Ag, Cu and so on under well-controlled Ar stream. Coating with some organic compounds allows these metal nanoparticles to be dispersed in an appropriate organic solvent. These dispersed nanoparticles in organic solvents are now commercially available.

So far as passivated Ag nanoparticles concerns, there is known a convenient one-pot preparation method *(14)*. Silver salt of carbonic acid such as myristate is heated at about 620 K without any solvent. Pyrolytically reduced Ag atoms then self-assemble to form metal clusters of about 5 nm in diameter, which are surrounded by remaining Ag myristate to form organo-passivated Ag nanoparticles with yield of 70 %. This type of self-organization may be called reaction-induced self-assembly.

One of the interesting features of metal nanoparticles is their melting at very low temperature. Melting point of bulk Au and Ag is 1,337 K and 1,235 K, respectively, but their organopassivated nanoparticles with diameter less than 10 nm melt at below or much lower than 500 K. If line-fused, a resulting metal-color line shows good electric conductivity even though the nanoparticles are initially passivated by organic compounds. By use of this low-temperature melting nature, we can draw electric circuits on many practical substrates including polymers and papers. Metal nanoparticles are therefore expected to bring about a technological innovation in the near future.

The most practical way to draw lines and dots by nanoparticle-ink is to use a PC-controlled ink jet printer. However, we can propose an alternative method based on a new concept of self-organization, i.e., dissipative structure-assisted self-assembly.

Dissipative Structure-Assisted Self-Assembly of Metal Nanoparticles

Strategy to Realize Hierarchic Structure

Suppose we want to obtain a dot or line-shaped pattern composed of metal nanoparticles, the characteristic wavelength of which is the order of μm. It is a typical problem to realize a hierarchic structure, as the wavelengths in the resulting structure are different by 3 orders of magnitude. So we attribute the μm-scale pattern formation of dots or lines to one layer of hierarchy (super-layer) and aggregation of nano-size particles into a shape of dot or line attributes to another layer (sub-layer), and apply different strategies to different layers.

Aggregation of nanoparticles is an issue of self-assembly in the sub-layer. Once the boundary conditions are fixed, aggregation will proceed to reach to the equilibrium state. The pattern formation in the super-layer simultaneously provides the initial and boundary conditions for the aggregation process in the sub-layers. In order to obtain in the super-layer a characteristic wavelength that is sufficiently larger than the size of the components, we introduce a dissipative structure principle. Such a dissipative pattern is known in the dewetting process of a dilute polymer solution *(15)*. Behind the receding front of polymer solution is left an array of polymer dots with a few μm in diameter and with lateral spacing of about 10 μm. If nanoparticles are left within polymer dots that still contain solvent to some extent, the nanoparticles must start to aggregate in accordance with further evaporation of solvent (Figure 4).

24

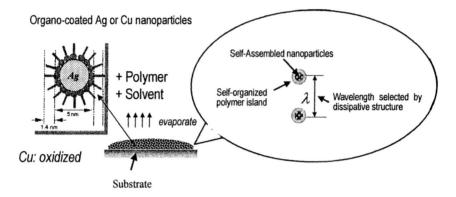

Figure 4. Experimental design based on the new concept of self-organization to obtain a hierarchic dot or stripe pattern composed of metal nanoparticles. Here polymer is used for constructing the super-layer structure with the wavelength λ that is chosen by dewetting instability. The patterned polymer islands simultaneously provide the initial and boundary conditions of sub-layer, i.e., the conditions for self-assembly of nanoparticles.

Experimental

Polystyrene (M_w: 166,400-187,200) was purchased from Wako Pure Chemicals. Toluene of reagent grade was uses without purification. Organo-passivated Cu and Ag nanoparticles dispersed in toluene were gifts from Harima Chemicals Inc. (concentrations of metals: < 40 wt%) *(16)*. Diameter of metal core of nanoparticles is about 5 nm and total diameter is about 7-8 nm.

A thin toluene solution of metal nanoparticles and polystyrene (1:1 in weight) was evaporated on a clean glass plate at room temperature (299 K). The glass was kept vertically (70 degrees) till toluene evaporated completely.

The self-assembled hierarchic patterns of nanoparticles were observed by optical microscope (Olympus STM, object lens ULWD MSPlan50).

Results and Discussion

A typical example of dissipative structure-assisted self-assembly of nanoparticles is shown in Figure 5. The dark dots and lines are aggregates of organopassivated Cu nanoparticles, and the light surroundings are polystyrene matrices. The dewetting process determines first the wavelength of spatial pattern (frozen dissipative structure) of polymer matrices that are isolated one another, then followed by the self-assembly process of nanoparticles in each polymer island matrix. Apparently, the aggregate of nanoparticles locate at the center of each polymer matrix. It suggests that gradual evaporation of solvent caused precipitation and aggregation (phase separation) of hydrocarbon-passivated nanoparticles in the matrix of polystyrene-toluene mixture island. By choosing other combinations of passivating organic layer and polymer, we may obtain homogeneous dispersion of nanoparticles in polymer matrices.

In Figure 5, there exist several line-shaped structures. These lines run along the moving direction of receding front. One may notice that some lines look like they are going to split. This observation suggests a possible route of producing a dot pattern via instability of line-shaped structures.

Basically, line-shaped structures can be organized in both directions, parallel and normal to the direction of receding front, depending on experimental conditions. Interestingly, we can occasionally obtain a specimen where polymer lines run orthogonal with different wavelengths in the dewetting process of polymer solution *(17)*. Similar patterns with two orthogonal waves has been observed in a crystallization process of L-ascorbic acid from a sparsely distributed methanol solution as well.

The dewetting pattern of polymer on glass is highly sensitive to a number of experimental parameters (the primary factors): the initial concentration of polymer solution, the thickness of solution, solvents, temperature, the conditions of glass surface, humidity in air, and others. As the pattern formation via dewetting is a physical process and is observable in many other polymeric systems, we can reasonably expect that these experimental factors are included

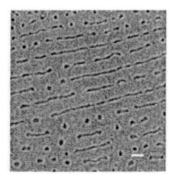

Figure 5. Self-assembly of Cu-nanoparticles (passivated by hydrocarbons) in polystyrene matrices. A thin toluene solution of 1.0 mg/mL Cu-nanoparticles and 0.95mg/mL polystyrene was evaporated in air on a glass plate held vertically at 299 K. The white bar indicates 2 μm.

into the secondary factors such as the viscosity and the velocity of receding front, by which the cubic shape of the receding front is described. When the shape is known, the controllability of dewetting process may be improved based on the knowledge on the instability of receding front. The authors therefore consider this modeling approach important, and it is in progress in our laboratory in parallel to the laboratory experiments.

Further Possibilities of Dissipative Structure-Assisted Self-Assembly in Polymeric Systems

Once we notice and recognize the dissipative structure-assisted self-assembly, we can find many examples in literature, typically in the field of materials sciences and molecular biology. Listed below are some examples related to polymeric systems.

1. The spin-mode frontal polymerization resulting in a helicoidal structure *(18)*.
2. A record of the traveling BZ spiral as a spiral deposition of AgBr in a gel membrane *(8)*, though it requires an additional assistance of experimentalists.
3. Photoinduced phase separation in a polymer mixture that results in a Turing-type structure *(9)*.
4. Periodic radical polymerization in the BZ media *(19)*.
5. Microscale patterns induced by laser ablation *(20)*.
6. Clustering of receptor molecules in a biological cell *(21)*.
7. Pattern formation in most of convecting systems.
8. Treadmilling of actin microfilaments, if controlled by coexisting dissipative structures.

This list is not enough; nevertheless, it will be a guide for readers to understand correctly the potentials of dissipative structure-assisted self-assembly mechanism, and to design new experimental systems and practical application of polymers.

Conclusions

Self-assembly and dissipative structure formation are complementary self-organizing processes. Making better use of these two principles, we designed and realized hierarchic structure composed of metal nanoparticles. The present comprehensive idea of self-organization is a useful concept to design and realize a hierarchic order, especially in nano-scale technology, and to understand biological systems.

Acknowledgments

TY thanks Dr. Olaf Karthaus and Dr. Masatsugu Shimomura for stimulative discussion. This research has been supported partly by the Shorai Foundation for Science and Technology.

References

1. Schrödinger, E. *What is Life? The Physical Aspect of the Living Cell*; Cambridge University Press: Cambridge, 1944.
2. Nicolis, G.; Prigogine, I. *Self-Organization in Non-Equilibrium Systems*; John Wiley & Sons, Inc.: New York, 1977.
3. Whitesides, G. M.; Grzybowski, B. *Science*, **2002**, *295*, 2418-2421.
4. Hollingsworth, M. D. *Science*, **2002**, *295*, 2410-2413.
5. Aliev, R.R.; Davydov, V. A.; Ohmori, T., Nakaiwa, M.; Yamaguchi, T. *J. Phys. Chem. A*, **1997**, *101*, 1313-1316.
6. Turing, A. *Phyl. Trans. R. Soc. London* **1952**, *B327*, 37-72.
7. Nishiura, M. *Far-from-Equilibrium Dynamics*; American Mathematical Society, 2002.
8. Köhler, J. M.; Müller, S.C. *J. Phys. Chem.*, **1995**, *99*, 980-983.
9. Tran-Cong, Q.; Kawai, J.; Endoh, K. *Chaos*, **1999**, *9*, 298-307.
10. Pearson, J. E. *Science*, **1993**, *261*, 189-192.
11. Mullins, W. W.; Sekerka, R.F. *J. Appl. Phys.* **1964**, *35*, 444-451.
12. Lefever, R. *Workshop on Dissipative Structures and Non-Equilibrium Assisted Self-Assembly*, Foundation des Treilles, Tourtour, 2000.
13. Yamaguchi, T. *Chem. Chem. Ind.*, **2001**, *54*, 1363-1367 (in Japanese).
14. Nagasawa, H; Maruyama, M.; Komatsu, T.; Isoda, S.; Kobayashi, T. *Phys. Stat. Sol.*, **2002**, *191*, 67-76.
15. Karthaus, O.; Ijiro, K.; Shimomura, M. *Chem. Lett.*, **1996**, 821-822.
16. See http://www.harima.co.jp/products/electronics/.
17. Shimomura, M. Private communication.
18. Pojman, J. A.; Masere, J.; Petretto, E.; Rustici, M.; Huh, D. S.; Kim, M. S.; Volpert, V. *Chaos*, **2002**, *12*, 56-65.
19. Washington, R. P.; Misra, G. P.; West, W. W.; Pojman, J. A. *J. Am. Chem. Soc.* **1999**, *121*, 7373-7380.
20. Niino, H.; Ihlemann, J.; Ono, S.; Yabe, A. *Macromol. Symp.*, **2000**, *160*, 159-165.
21. Bray, D. *Nature*, **1998**, *393*, 85-88.

Gels

Chapter 3

Synchronization of Self-Oscillation in Polymer Chains and the Cross-Linked Network

Ryo Yoshida[1,*], Takamasa Sakai[1], Shoji Ito[2], and Tomohiko Yamaguchi[2]

[1]Department of Materials Engineering, Graduate School of Engineering, The University of Tokyo, Tokyo 113–8656, Japan
[2]Nanotechnology Research Institute, National Institute of Advanced Industrial Science and Technology, AIST Central 5–2, 1–1–1 Higashi, Tsukuba 305–8565, Japan

Novel biomimetic gel that undergoes autonomous swelling-deswelling oscillations without on-off switching of external stimuli was developed. The mechanical oscillation was produced via the Belousov-Zhabotinsky reaction. The gel consists of the cross-linked poly(N-isopropylacrylamide) to which ruthenium tris(2,2'-bipyridine) was covalently bonded. The redox oscillation of the catalyst site is converted into the mechanical oscillation of the polymer network. The oscillating behaviors were controlled by changing the substrate concentration, gel size or geometry, photo-illumination, etc. For the analysis of synchronization process in the gel, linear polymer chain and gel particles with submicrometer size were prepared. For linear polymer, soluble-insoluble changes were realized. Effects of polymerization and crosslinking were investigated through the analysis of oscillation for these systems.

Stimuli-responsive polymer gels undergo either swelling or deswelling transition when varying external conditions surrounding the gels such as a change in solvent composition (*1, 2*), pH (*3, 4*), temperature (*5, 6*), electric field (*7, 8*), specific chemicals (*9-11*), etc. For example, thermo-sensitive hydrogels consisting of N-isopropylacrylamide (NIPAAm) swell by cooling and deswell by heating (*5, 6*). Many kinds of stimuli responsive gels have been extensively investigated and their ability to swell and deswell according to conditions makes them an interesting proposition for use in intelligent materials (*12-17*).

In these systems utilizing stimuli-responsive polymers, the response of polymer is temporary; that is, the polymer provides only one unique action of either expanding or collapsing toward a stable equilibrium state. Therefore the on-off switching of external stimuli is essential to instigate the action of the polymer. On the other hand, many physiological systems maintain rhythmical oscillations under constant environmental conditions, and act in a dynamic nonequilibrium state, as represented by the autonomic heartbeat, brain waves, periodic hormone secretion. If such self-oscillation could be achieved for gels, possibilities would emerge for new biomimetic intelligent materials that exhibit rhythmical motion. Recently, we have developed such a self-oscillating gel (*18-23*). It spontaneously exhibits cyclic swelling and deswelling under constant conditions, requiring no switching of external stimuli. Its action is similar to that of a beating heart muscle. In this paper, the self-oscillating behaviors of the gel have been discussed.

Design of Self-Oscillating Gel

The mechanical oscillation is driven by the Belousov-Zhabotinsky (BZ) reaction occurring in the gel (Figure 1). We prepared a copolymer gel which consists of N-isopropylacrylamide (NIPAAm) and ruthenium(II) tris(2,2'-bipyridine) (Ru(bpy)$_3$). Ru(bpy)$_3$, acting as a catalyst for the BZ reaction, is pendent to the polymer chains of NIPAAm. Homopolymer gels of NIPAAm have thermosensitivity and undergo an abrupt volume-collapse (phase transition) when heated at around 32°C (5,6). The oxidation of the Ru(bpy)$_3$ moiety caused not only an increase in the swelling degree of the gel, but also a rise in the transition temperature. These characteristics may be interpreted by considering an increase in hydrophilicity of the polymer chains due to the oxidation of Ru(II) to Ru(III) in the Ru(bpy)$_3$ moiety. As a result, we may expect that our gel undergoes a cyclic swelling-deswelling alteration when the Ru(bpy)$_3$ moiety is periodically oxidized and reduced under constant temperature.

32

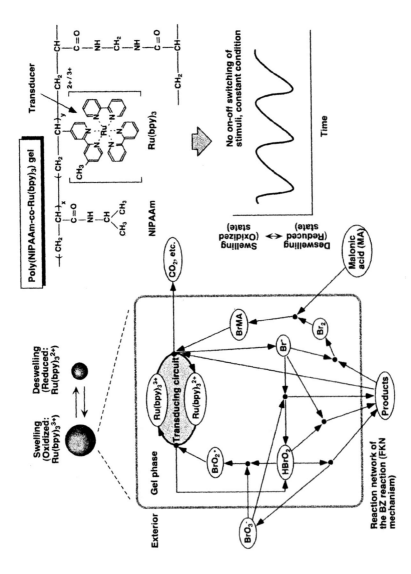

Figure 1. Self-oscillation of poly(NIPAAm-co-Ru(bpy)₃) gel coupled with the Belousov-Zhabotinsky reaction.

Self-Oscillation of the Miniature Bulk Gel

The poly(NIPAAm-co-Ru(bpy)$_3$) gel was cut into a cubic shape (each length of about 0.5mm) in pure water, and then immersed into an aqueous solution containing malonic acid (MA), sodium bromate (NaBrO$_3$), and nitric acid (HNO$_3$) at constant temperature (20°C). This outer solution comprised the reactants of the BZ reaction, with the exception of the catalyst. Therefore the redox oscillation does not take place in this solution. However, as it penetrates into the gel, the BZ reaction is induced within the gel by the Ru(bpy)$_3$ copolymerized as a catalyst on the polymer chains. Under reaction, the Ru(bpy)$_3$ in the gel network periodically changes between 2+ and 3+ states. In the miniature gel whose size is smaller enough than the wavelength of chemical wave (typically several mm), the redox change of ruthenium catalyst can be regarded to occur homogeneously without pattern formation. We observed the oscillation behavior under a microscope equipped with a CCD camera and video recorder. Color changes of the gel accompanied with redox oscillations (orange: reduced state, light green: the oxidized state) were converted to 8-bit grayscale changes (dark: reduced, light: oxidized) by image processing. Due to the redox oscillation of the immobilized Ru(bpy)$_3$, mechanical swelling-deswelling oscillation of the gel autonomously occurs with the same period as for the redox oscillation (Figure 2). The volume change is isotropic and the gel beats as a whole, like a heart muscle cell. The chemical and mechanical oscillations are synchronized without a phase difference (i.e., the gel exhibits swelling during the oxidized state and deswelling during the reduced state).

In order to enhance the amplitude of swelling-deswelling oscillations of the gel, we attempted to change the period and amplitude of the redox oscillation by varying the initial concentration of substrates. It is a general tendency that the oscillation period increases with the decrease in concentration of substrates. For the bulk solution consisting of MA, NaBrO$_3$, HNO$_3$ and Ru(bpy)$_3$Cl$_2$, we obtained the following empirical relations between the period (T [s]) and initial molar concentration of substrates: T = 2.97 [MA]$^{-0.414}$[NaBrO$_3$]$^{-0.796}$[HNO$_3$]$^{-0.743}$. For the self-oscillating gel, the empirical relations was different from the bulk solution system as follows : T = 60.3 [MA]$^{-0.155}$[NaBrO$_3$]$^{-0.436}$[HNO$_3$]$^{0.469}$. The reason may be as follows. (1) In the case of the miniature gel, the dilution of intermediates from the gel into the surrounding aqueous phase must be more remarkable. The dilution effect, especially that for the activator (HBrO$_2$), leads to an increase in the period of chemical oscillations. (2) In addition, concentration change of substrates or products within the gel phase resulting from the swelling-deswelling oscillations may have some effects on the chemical oscillations (i.e., feedback effect).

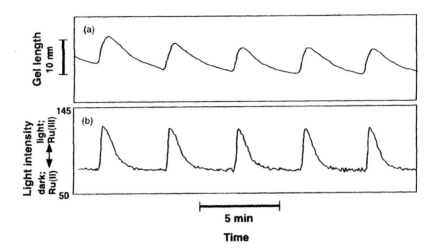

Figure 2. (a) Periodical redox changes of the miniature cubic poly(NIPAAm-co-Ru(bpy)₃) gel and (b) the swelling-deswelling oscillation at 20°C. Transmitted light intensity is expressed as an 8-bit grayscale value. Outer solution: [MA] = 62.5mM; [NaBrO₃] = 84mM; [HNO₃] = 0.6M.
(Reproduced from reference 22. Copyright 2000 American Chemical Society.)

The variation in chemical oscillation leads to a change in the swelling-deswelling oscillation: i.e., the swelling-deswelling amplitude (the change in gel length, Δd) increases with an increase in the period and amplitude of the redox changes. Empirically, the relation between Δd [μm] and the substrate concentrations was expressed as : $\Delta d = 2.38[MA]^{0.392}[NaBrO_3]^{0.059}[HNO_3]^{0.764}$.

As a result, it is apparent that the swelling-deswelling amplitude of the gel is controllable by changing the initial concentration of substrates. So far, the swelling-deswelling amplitude with ca. 20% to the initial gel size was obtained as a maximum value. It is worth noting that the waveform of redox changes deforms to rectangular shape with a plateau period when the amplitude of swelling-deswelling oscillation increased (Fiugre 3). From this result, it is supposed that not only energy transformation from chemical to mechanical change, but also feedback mechanism from mechanical to chemical change acts in the synchronization process.

Peristaltic Motion of the Rectangular Gel

Figure 4 shows the rectangular piece of the gel (ca. 1mm × 1mm × 20mm) which was immersed in the aqueous solution containing the three reactants of the BZ reaction. The chemical waves propagate in the gel at a constant speed in the

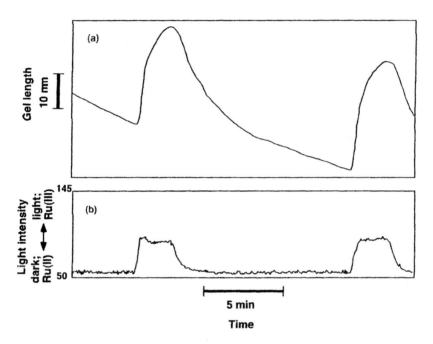

Figure 3. (a) Periodical redox changes of the miniature cubic poly(NIPAAm-co-Ru(bpy)₃) gel and (b) the swelling-deswelling oscillation at 20°C. Transmitted light intensity is expressed as an 8-bit grayscale value. Outer solution: [MA] = 62.5mM; [NaBrO₃] = 84mM; [HNO₃] = 0.894M.
(Reproduced from reference 22. Copyright 2000 American Chemical Society.)

direction of the gel length. Considering the dark (Ru(II)) and light (Ru(III)) zones represent simply the shrunken and swollen parts respectively, the locally swollen and shrunken parts move with the chemical wave, like peristaltic motion of living worms. The propagation of the chemical wave makes the free end of the gel move back and forth at a rate corresponding to the wave propagation speed. As a result, the total length of the gel periodically changes. It was demonstrated by the mathematical model simulations that the change in the overall gel length is equivalent to that in the remainder of gel length devided by the wavelength, because the swelling and the deswelling cancel each other per one period of oscillations under steady oscillating conditions (20).

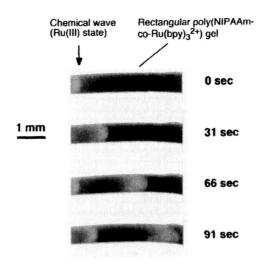

Figure 4. Propagation of chemical wave in the rectangular poly(NIPAAm-co-Ru(bpy)₃) gel.

It is well known that the period of oscillation is affected by light illumination for the $Ru(bpy)_3^{2+}$-catalysed BZ reaction. The excited state of the catalyst $(Ru(bpy)_3^{2+*})$ causes new reaction process: production of activator (reaction (1)), or production of inhibitor (reaction (2)), which depends on the solute compositions (24).

$$Ru(bpy)_3^{2+*} + Ru(bpy)_3^{2+} + BrO_3^- + 3H^+ \rightarrow HBrO_2 + 2Ru(bpy)_3^{3+} + H_2O \quad (1)$$
$$Ru(bpy)_3^{2+*} + BrMA + H^+ \rightarrow Br^- + Ru(bpy)_3^{3+} + products \quad (2)$$

Therefore, (i) we can intentionally make a pacemaker with a desired period (or wavelength) by local illumination of laser beam on the gel, or (ii) we can change the period (or wavelength) by local illumination on the pacemaker which has already existed in the gel.

In the rectangular gel, the corner often becomes a pacemaker from which chemical waves start to propagate. Therefore, the self-oscillating behaviors of the gel can be controlled by irradiating laser light locally to the pacemaker site of gels. Figure 5 shows the effect of laser irradiation (488nm) on the pacemaker under the condition that photo-illumination produces activator. The size of the pacemaker was controlled by the illuminating beam, whose diameter was determined by a pinhole placed in the light path. It was found that the wavelength of traveling waves in the gel decreased as the size of pacemaker increased. The results gave a good agreement with a theoretical model simulation. This result means that we can control the macroscopic swelling-deswelling hebavior of the gel by local perturbation, i.e., small signal can be amplified to macroscopic change.

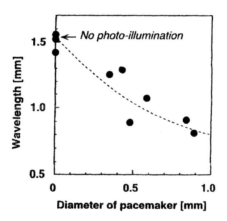

Figure 5. Relation between the diameter of pacemaker and the wavelength of chemical waves.

Self-Oscialltion of Polymer Chains with Rhythmical Soluble-Insoluble Changes

In the self-oscillating gel, redox changes of Ru(bpy)₃ catalyst are converted to confomational changes of polymer chain by polymerization. The conformational changes are amplified to macroscopic swelling-deswelling changes of polymer network by crosslinking. Further, when the gel size is larger than chemical wavelength, the chemical wave propagates in the gel by coupling with diffusion. Then peristaltic motion of the gel is created. In these manners, hierarchical synchronization process exists in the self-oscillating gel. Our interests are to clarify the polymerization effect of catalyst on the oscillating behavior of the BZ reaction, as well as the effect of crosslinking the polymer chains on the synchronization of each polymer's oscillation. For this purpose, firstly, we synthesized linear poly(NIPAAm-co-Ru(bpy)₃) and investigated self-oscillating behavior of polymer chain through the analysis of transmittance changes of polymer solution (25). Then the gel particles with submicronmeter size were prepared. By comparing the oscillating behaviors between them, the effect of polymerization and crosslinking were discussed.

Figure 6 shows the transmittance changes of poly(NIPAAm-co-Ru(bpy)₃) (5wt% and 10wt% Ru(bpy)₃) solutions as a function of temperature under the different conditions of reduced Ru(II) sate and oxidized Ru(III) state. The wavelength (570nm) at isosbestic point of reduced and oxidized states was used to detect the optical transmittance changes based on soluble-insoluble changes of the polymer, not on the redox changes of Ru(bpy)₃ moiety. The transmittance suddenly decreases as temperature increases, demonstrating the lower critical solution temperature (LCST). When the Ru(bpy)₃ site is kept in an oxidized state, the LCST shifts higher than that of the reduced state. The rise in the LCST by oxidation is due to an increase in hydrophilicity of the polymer by the charge

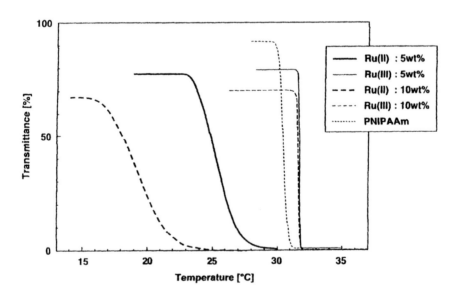

Figure 6. Temperature dependence of optical transmittance for poly(NIPAAm-co-Ru(bpy)₃) solutions under the different conditions of reduced Ru(II) state (in Ce(III) solution) and oxidized Ru(III) state (in Ce(IV) solution).

increase of the catalyst. The difference of the LCST between reduced and oxidized states becomes larger as the Ru(bpy)₃ content increases in the polymer.

In reduced state, the effect of hydrophobicity of bipyridine ligand surrounding ruthenium ion would be predominant over the ionization effect and make the polymer more hydrophobic than PNIPAAm. Due to the hydrophobic interaction, the sharpness of transmittance change may become duller in reduced state, and the LCST decreases more significantly with the increase in Ru(bpy)₃ content. In oxidized Ru(III) state, on the other hand, the LCST of the copolymer is slightly higher than that of PNIPAAm. As charge number increases, the contribution of ionic effect increases, which makes the polymer more hydrophilic than PNIPAAm.

From the deviation of the LCST between Ru(II) and Ru(III) states, we may expect that the polymer undergoes periodical soluble-insoluble changes when the Ru(bpy)₃ moiety is oxidized and reduced periodically by the BZ reaction at constant temperature. Synchronized with the periodical changes between Ru(II) and Ru(III) states of the Ru(bpy)₃ site, the polymer becomes hydrophobic and hydrophilic, and exhibits cyclic soluble-insoluble changes. These periodic changes of polymer chains can be easily observed as cyclic transparent and opaque changes for the polymer solution with color changes due to the redox oscillation of the catalyst. Figure 7 shows the oscillation profiles of transmittance for the polymer solution at constant temperatures. As temperature increases, the amplitude in transmittance increases and the period becomes short. It is a general tendency that the oscillation period of the BZ reaction decreases as temperature increases, following the Arrhenius equation (26).

As observed in Figure 6, the degree of difference in transmittance between reduced and oxidized states depends on temperature because the transmittance in reduced state drops more gradually over wide temperature range while drops abruptly in oxidized state. Consequently, the difference in transmittance between two states becomes large at high temperature. This results in the large amplitude of oscillations at high temperature in Figure 7(a). Figure 7(b) shows the oscillating behavior of the polymer with higher Ru(bpy)₃ content. The amplitude increases at each temperature. This result is attributed to the large difference of LCST between reduced and oxidized states of high Ru(bpy)₃ content (see Figure 6). It is suggested that the hydrophilic-hydrophobic changes of polymer becomes more remarkable due to an increase in molar content of redox site.

We have investigated dependence of the oscillation period on initial substrate concentration for three systems; (i) the conventional BZ solution using non-polymerized catalyst, (ii) the polymer solution using polymerized catalyst by NIPAAm, and (iii) the suspension of submicron-sized gel beads, i.e., the crosslinked polymer network of the polymerized catalyst. The empirical equation, $T = a[MA]^{-b}[NaBrO_3]^{-c}[HNO_3]^{-d}$, was obtained for these three systems. It was found that the concentration dependence of period for the non-polymerized solution (each constants in the equation are a=2.97, b=0.414, c=0.794 and d=0.743, respectively) was partly similar to the polymer solution

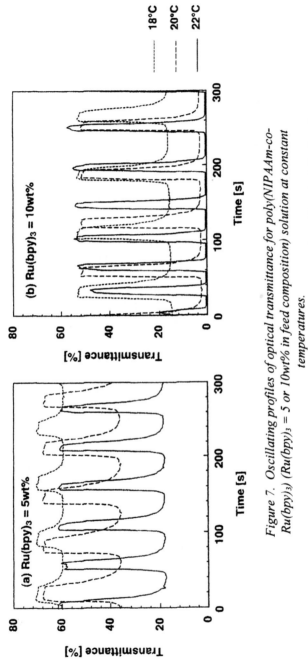

Figure 7. Oscillating profiles of optical transmittance for poly(NIPAAm-co-Ru(bpy)₃) (Ru(bpy)₃ = 5 or 10wt% in feed composition) solution at constant temperatures.

(Reproduced from reference 25. Copyright 2002 American Chemical Society.)

(a=2.97, b=0.413, c=0.934 and d=0.567), but largely different from the gel beads suspension (a=5.75, b=0.506, c=0.667 and d=0.478). Compared with the period under the same substrate concentrations, the period increased as the following order; (i) < (ii) < (iii).

The reason for the increases in perod can be considered as follows. In the poly(NIPAAm-co-Ru(bpy)$_3$), charge site is fixed on the polymer chain. Due to the fixation, the increase in electrostatic repulsion between the charge sites may be suppressed. As a result, change to oxidation state is restrained. This leads to longer duration of reduced state and therefore elongation of oscillation period. In the case of crosslinked polymer network (gel beads), polymer chains are constrained to behave cooperatively. And also, diffusion limitation of substrate from the outer solution into gel phase will take place. This will result in decreasing effective concentration inside the gel phase and elongation of oscillation period. Detail mechanisms are still under investigation.

Ciliary Motion Actuator Using Self-Oscillating Gel

One of the promising fields of the MEMS is micro actuator array or distributed actuator systems. The actuators, that have a very simple actuation motion such as up and down motion, are arranged in an array form. If their motions are random, no work is extracted from this array. However, by controlling them to operate in a certain order, they can generate work as a one system. One of the typical examples of this kind of actuation array is a ciliary motion micro actuator array. There have been many reports to realize it. Although various actuation principles have been proposed, all the previous works based on the same concept that the motion of actuators were controlled by external signals. If the self-oscillating gel plate with micro projection structure array on top is realized, it is expected that the chemical wave propagates and creates dynamic rhythmic motion of the micro projection structure array. This is the structure of proposed new ciliary motion array that exhibits spontaneous dynamic propagating oscillation.

The gel plate with micro projection array was fabricated by molding technique (27). First, the moving mask deep-X-ray lithography technique was utilized to fabricate the PMMA plate with truncated conical shape microstructure array. This step was followed by the evaporation of Au seed layer and subsequent electroplating of nickel to form the metal mold structure. Then, a PDMS mold structure was duplicated from the Ni metal mold structure and utilized for gel molding. The formation of gel was carried out by vacuum injection molding technique. The structure with the height of 300 μm and bottom diameter of 100 μm were successfully fabricated by the proposed process (Figure 8). The propagation of chemical reaction wave and dynamic rhythmic motion of the micro projection array were confirmed by chemical wave observation and displacement measurements (27). The feasibility of the new

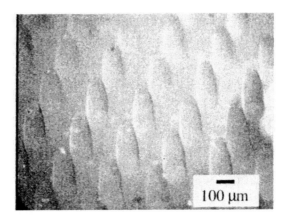

Figure 8 The micro projection structure array on the gel surface fabricated by X-ray lithography technique.

concept of the ciliary motion actuator made of self-oscillating polymer gel was successfully confirmed.

Conclusions

Novel biomimetic polymer gels with self-oscillating function have been developed. The gel has a cyclic reaction network in itself and generates periodic mechanical energy from the chemical energy of the BZ reaction. The self-oscillating behavior can be controlled by changing the reaction condition or geometric design of the gel. The self-oscillating gel may be useful in a number of important applications such as self-walking (auto-mobile) actuators or micropumps with autonomous beating or peristaltic motion, etc. Research into the mechanism of the oscillating behavior as well as the practical applications continues.

References

1. Dusek, K. ed. *Responsive Gels: Volume Transitions II,* Berlin: Springer-Verlag, 1993
2. Tanaka, T.; Fillmore, D.; Sun, S.T.; Nishio, I.; Swislow, G.; Shah, A. *Phys. Rev. Lett.,* **1980**, 45, 1636-1639.
3. Kawasaki, H.; Sasaki, S.; Maeda, H. *J. Chem. Phys.* **1997**, 101, 5089-5093.
4. Siegel, R.A.; Falamarzian, M.; Firestone, B.A.; Moxley, B.C. *J. Controlled Release,* **1998**, 8, 179-182.

5. Hirokawa, Y.; Tanaka, T. *J. Chem. Phys.* **1984**, 81, 6379-6380.
6. Yoshida, R.; Uchida, K.; Kaneko, Y.; Sakai, K.; Kikuchi, A.; Sakurai, Y.; Okano, T. *Nature* **1995**, 374, 240-242.
7. Tanaka, T., Nishio, I., Sun, S.-T.; Ueno-Nishio, S. *Science*, **1982**, 218, 467-469.
8. Osada, Y.; Okuzaki, H.; Hori, H. *Nature* **1992**, 355, 242-244.
9. Oya, T.; Enoki, T.; Grosberg, A.Y.; Masamune, S.; Sakiyama, T.; Takeoka, Y.; Tanaka, K.; Wang, G.; Yilmaz, Y.; Feld, M.S.; Dasari, R.; Tanaka, T. *Science*, **1999**, 286, 1543-1545.
10. Kataoka, K.; Miyazaki, H.; Bunya, M.; Okano, T.; Sakurai, Y. *J. Am. Chem. Soc.*, **1998**, 120, 12694-12695.
11. Miyata, T.; Asami, N.; Uragami, T. *Nature*, **1999**, 399, 766-769.
12. Okano, T. ed. *Biorelated polymers and gels: controlled release and applications in biomedical engineering*, Academic Press: Boston, 1998.
13. Dong, L.C.; Hoffman, A.S. *J. Controlled Release*, **1991**, 15, 141-152.
14. Yoshida, R.; Sakai, K.; Okano, T.; Sakurai, Y. *Advanced Drug Delivery Reviews*, **1993**, 11, 85-108.
15. Holtz, J.H.; Asher, S.A. *Nature*, **1997**, 389, 829-832.
16. Kiser, P.F.; Wilson, G.; Needham, D. *Nature*, **1998**, 394, 459-462.
17. Beebe, D.J.; Moore, J.S.; Bauer, J.M.; Yu, Q.; Liu, R.H.; Devadoss, C.; Jo, B.H. *Nature*, **2000**, 404, 588-590.
18. Yoshida, R.; Takahashi, T.; Yamaguchi, T.; Ichijo, H. *J. Am. Chem. Soc.*, **1996**, 118, 5134-5135.
19. Yoshida, R.; Takahashi, T.; Yamaguchi, T; Ichijo, H. *Adv. Mater.* **1997**, 9, 175-178.
20. Yoshida, R.; Kokufuta, E.; Yamaguchi, T. *CHAOS* **1999**, 9, 260-266.
21. Yoshida, R.; Onodera, S.; Yamaguchi, T.; Kokufuta, E. *J. Phys. Chem. A*, **1999**, 103, 8573-8578.
22. Yoshida, R.; Tanaka, M.; Onodera, S.; Yamaguchi, T.; Kokufuta, E. *J. Phys. Chem. A*, **2000**, 104, 7549-7555.
23. Yoshida, R.; Otoshi, G.; Yamaguchi, T.; Kokufuta, E. *J. Phys. Chem. A*, **2001**, 105, 3667-3672.
24. Amemiya, T.; Ohmori, T.; Yamaguchi, T. *J. Phys. Chem. A*, **2000**, 104, 336-344.
25. Yoshida, R.; Sakai, T.; Ito, S.; Yamaguchi, T. *J. Am. Chem. Soc.*, **2002**, 124, 8095-8098.
26. Ruoff, P. *Physica D*, **1995**, 84, 204-211.
27. Tabata, O.; Hirasawa, H.; Aoki, S.; Yoshida, R.; Kokufuta, E. *Sensor and Actuators A*, **2002**, 95, 234-238.

Chapter 4

A Model for a Hydrogel–Enzyme Chemomechanical Oscillator

Anish P. Dhanarajan[1], Jon Urban[1], and Ronald A. Siegel[1,2,*]

Departments of [1]Biomedical Engineering and [2]Pharmaceutics, University of Minnesota, Minneapolis, MN 55455

A glucose-fueled hydrogel/enzyme construct, which may be useful for the autonomous, rhythmic delivery of peptides such as gonadotropin releasing hormone (GnRH), is under investigation. In the present work, an illustrative mathematical model of this system is presented. The model combines aspects of the physical chemistry of hydrophobic polyelectrolyte hydrogels and of chemical kinetics. With this model we explore factors affecting the system's capacity to oscillate, and how certain characteristics of the oscillations are affected by design parameters.

Introduction

Hydrogels are crosslinked polymer networks that absorb water. In a static environment, these materials swell to an extent that is determined, at equilibrium, by a balance of thermodynamic forces arising from the interspersion of polymer with water and the elastic response to stretch or compression of polymer chains away from their most favorable configurations (*1,2*). Physical or chemical changes in the environment lead to changes in hydrogel swelling, the kinetics of which depend on hydrogel dimensions and permeability to water.

Polyelectrolyte hydrogels contain fixed, ionizable pendant groups, which interact electrostatically with each other and with microions in the interstitial water. These interactions give rise to a Donnan osmotic pressure (*1,3,4*), which provides an extra swelling force. The Donnan swelling force is determined, in turn, by the density of fixed ionized groups and the concentrations of microions inside the hydrogel. Both of these factors may be controlled by pH, ionic strength, and the presence of specific molecules in the environment. Polyelectrolyte hydrogels are therefore attractive as components of chemical sensors and chemomechanical actuators. For example, devices containing glucose-sensitive hydrogels can measure blood glucose levels and actuate insulin delivery in response to changes in blood glucose level (*5,6*).

Certain polyelectrolyte hydrogels display first-order volume phase transitions, with hysteresis, in response to external stimuli (*2,7*). Hydrophobic polyelectrolyte hydrogels, in particular, may undergo discrete transitions in response to changes in external pH. Such hydrogels have been considered as chemically-sensitive mechanical switches (*8*).

Polyelectrolyte hydrogels have also been examined as components in chemomechanical oscillators, in which the fixed charge density and hence the swelling state of a polyelectrolyte hydrogel is coupled to the oscillations of a nonlinear chemical reaction. In one example, swelling of a pH-sensitive hydrogel was slaved to a chemical (peroxide-sulfite-hexaferrocyanate) pH-oscillator (*9*). In a second system, a responsive hydrogel containing catalytic redox groups participated in the generation of oscillations in a Belousov-Zhabotinsky (BZ) medium (*10*).

A potential application of oscillating hydrogels lies in the rhythmic delivery of hormones to patients with hormonal deficiencies. For example, both males and females who lack proper endogenous rhythmic secretion of gonadotropin releasing hormone (GnRH) are unable to sustain normal reproductive function (*11*). The previously described oscillating hydrogel systems cannot be used for this purpose, since they involve toxic chemicals. As a more biocompatible alternative, we have designed a hydrogel/enzyme-based chemomechanical oscillator that functions in the presence of a constant supply of endogenous glucose (*11,12*).

In this article we first describe the experimental hydrogel/enzyme oscillator system, and display some experimental results. We then develop a mathematical model for this system, based on a combination of physicochemical theories used to describe hydrogel swelling, molecular transport through hydrogels, and the dynamics of chemical species outside the hydrogel. The purpose of such a model is to elucidate the influence of various design parameters on oscillator performance and to guide experiments. The model is subjected to bifurcation analysis in order to identify conditions supporting oscillations, and simulations confirming the predictions of bifurcation analysis are presented. Finally, we discuss the limitations of the present model and outline future work.

Construction of the Hydrogel/Enzyme Oscillator

Detailed accounts of the experimental methods employed to produce glucose-driven chemomechanical oscillations have been published previously (*11,12*). We briefly describe them here to facilitate a better understanding of the model. The experimental setup, illustrated in Figure 1, consists of a side-by-side diffusion cell containing two 75 ml, 50 mM saline solutions. The pH-sensitive polyelectrolyte hydrogel membrane is clamped between the two cells. Glucose-containing saline solution flows through one of the cells, denoted by Cell I. The pH of the solution in this cell is maintained at 7.00 by means of a pH stat. The other cell, Cell II, contains the dissolved enzymes glucose oxidase and catalase, and a solid piece of marble. The pH of the solution in cell II is measured and recorded continuously by a pH meter interfaced with a digital recorder. The solutions in both cells are stirred at 600 rpm and temperature is maintained at 37° C by a water jacket.

The hydrogel membrane is a copolymer of N-isopropylacrylamide (NIPA) doped with methacrylic acid (MAA) and cross-linked with small amounts of ethylene glycol dimethacrylate (EGDMA). The prepolymer solution containing the monomers at the desired concentrations is poured between two glass plates separated by a 250 μm spacer. Polymerization is initiated using a combination of ammonium persulfate as free radical generator and tetraethylene-methylenediamine as accelerator.

Permeability of the NIPA/MAA membrane to glucose is controlled by the pH of the solution in Cell II. Diffusion studies performed using radiolabeled glucose have shown that glucose transport through the membrane is substantially attenuated when the pH of the solution in Cell II is lower than a threshold value (*8*). However, when pH in Cell II is raised beyond that threshold, membrane permeability to glucose is increased about tenfold. The permeability change is a discontinuous function of pH and exhibits hysteresis. This phenomenon is a direct result of a first order volume phase transition exhibited by the hydrogel in response to change in pH. Hysteresis is observed in numerous systems displaying first order phase transitions, but the effect may be enhanced in the present case by the globally connected nature of the crosslinked hydrogel (*13*).

Figure 1b displays examples of pH oscillations observed in Cell II, with glucose level and pH fixed in Cell I. Oscillations persist for about one week, and period increases steadily with time. This drift, along with the ultimate cessation of oscillations, is believed to follow from the slow buildup of gluconate ion in Cell II (see below), which acts as a buffering species (*11,12*). In the present work, we shall ignore these aspects, which are not relevant to the capacity of the system to initiate oscillations. A more comprehensive analysis of the system is needed to account for these phenomena.

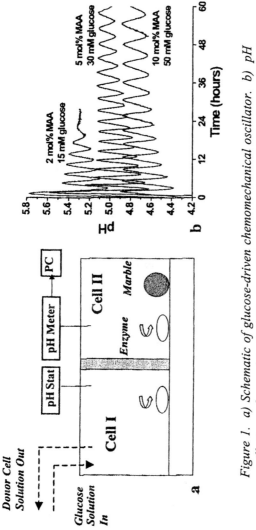

Figure 1. a) Schematic of glucose-driven chemomechanical oscillator. b) pH oscillations observed in Cell II at varying mole fractions of MAA incorporated into the hydrogel membrane, and varying glucose concentrations in Cell I. (Reproduced from reference 10. Copyright American Chemical Society.)

In Figure 1b it is shown that with reduced substitution of MAA in the hydrogel (10 to 5 mol%), there is an alkaline shift of the pH range over which oscillations occur, and the amplitude of oscillations decreases. Also, oscillations occur with exposure to a lower concentration of glucose in Cell I when MAA doping is decreased. With further decrease in MAA (2 mol%), oscillations become damped, and ultimately disappear at very low values of substitution (1 mol%, data not shown). One purpose of the model presented below is to explain these observations.

Mechanism of Chemomechanical Oscillations

As indicated previously, the hydrogel membrane's swelling and permeability to glucose exhibit a first-order transition, with hysteresis. In addition, delivery of glucose across the membrane leads to the production of hydrogen ions, which then control the membrane's swelling state, acting as a feedback. Since the combination of bistability (i.e. hysteresis) and feedback is a common feature of nonlinear electrical and chemical oscillators (14,15), these elements are also expected to play a mechanistic role in the present hydrogel/enzyme oscillator.

The mechanism of oscillations can be broken down into the following steps:

Step 1: Let the system start in its high permeability state, in which glucose readily diffuses from Cell I to Cell II. Glucose oxidase in Cell II converts glucose to gluconic acid, which dissociates into gluconate and hydrogen ions. This results in a drop in pH in Cell II (pH_{II}). As pH_{II} decreases, hydrogen ions diffuse into the hydrogel membrane and convert pendant carboxylates to the neutral carboxylic acid form.

Step 2: When a sufficient number of carboxylates have been neutralized, the membrane responds by switching to its collapsed, low permeability state. Consequently, glucose flux into Cell II and hydrogen ion production are attenuated, and pH_{II} levels off. Hydrogen ions in Cell II, in addition to diffusing into the membrane, are removed, in the presence of marble ($CaCO_3$), by a sequence of acid-base reactions (16):

$$H^+ + CaCO_3 \rightarrow Ca^{2+} + HCO_3^- \quad ; \quad H^+ + HCO_3^- \rightarrow H_2CO_3 \rightarrow CO_2 + H_2O$$

where the first of these reactions is heterogenous and rate limiting. This sequence of reactions constitutes a "shunt" pathway that accelerates the rate of pH change in Cell II. Without this shunt, pH oscillates at an unacceptably slow rate, and oscillations cease within a couple of periods, probably due to accumulation of gluconate ion, as discussed above (*11,12*).

During this step, pH_{II} increases. Also, there is significant deprotonation of pendant carboxylic acid groups in the membrane and release of hydrogen ions into Cell I, where they are carried away by flow. By this means, the hydrogel's fixed charge is gradually restored.

Step 3: When the hydrogel membrane becomes sufficiently charged, the membrane switches back to its swollen, high permeability state. The system is now poised at Step 1, and the cycle repeats.

Oscillations can only occur over a bounded range of glucose concentrations in Cell I. When glucose concentration is too low, the enzyme does not generate enough H^+ to convert the hydrogel from the swollen to the collapsed state. Conversely, when glucose concentration is too high, residual permeability of glucose through the collapsed membrane leads to sufficient production of hydrogen ions to maintain the membrane in the collapsed state.

Hysteresis in the membrane's swelling response facilitates oscillations. Without hysteresis, the system will find a steady state of intermediate permeability and glucose flux through the membrane.

Model Assumptions and Equations

The following assumptions were used to develop a mathematical model of swelling/deswelling oscillations:

1. Membrane properties, including the fixed charge concentration and swelling state, are assumed to be homogeneous in space, as are all ion concentrations inside the membrane. This simplifying assumption is made even though the membrane is subjected to a pH gradient.

2. Chemical potentials inside the membrane are determined by the Flory-Rehner-Donnan equation of state for one-dimensional swelling (*3*), since the clamped membrane's area, A, is constant. The polymer-solvent interaction parameter, χ, is a linear function of the polymer volume fraction, ϕ, i.e. $\chi = \chi_1 + \chi_2\phi$, as proposed by Hirotsu (*17*). Transport of water into and out of the hydrogel membrane causes its thickness, L, to change at a rate that is proportional to the difference in chemical potential of water between the membrane and Cell I and Cell II.

3. Enzymatic conversion of glucose to gluconic acid is assumed to be instantaneous. This assumption is valid when the concentration of enzyme in Cell II is sufficiently high that transport of glucose into Cell II is rate limiting regardless of the permeability of the membrane. Also, gluconate and bicarbonate are presumed to not perturb the state of the system.

4. Electroneutrality is maintained in the membrane at all times, primarily by Na^+ and Cl^-, whose concentrations adjust rapidly to changes in the fixed charge density of the hydrogel. The ratio of the concentration of a species with valence z inside the hydrogel to its concentration in the surrounding bath is given by the λ^z, where λ is the Donnan ratio (1,3,4). This holds true for Na^+ and H^+, both of which are present at higher concentration inside the membrane, and Cl^- and OH^-, which tend to be excluded from the membrane. The Donnan osmotic pressure difference is essentially due to partitioning of Na^+ and Cl^- between the inside of the hydrogel and the bath medium outside. The contribution of H^+ and OH^- to the Donnan osmotic pressure is neglected, since the concentrations of these ions are far below those of sodium and chloride.

5. The rate of change of fixed, negative charge concentration is controlled by the rate of transport of hydrogen ions, since a) they are the minority carrier species and b) they are subject to trapping by pendant carboxylates as they diffuse through the membrane (18).

6. Permeability of the membrane to glucose is expressed as $k_G{}^0 e^{-\beta\phi}$, as suggested by free volume theory (19). The parameters $k_G{}^0$ and β represent, respectively, the hypothetical glucose permeability with vanishing polymer concentration, and the sieving effect of the polymer, which depends on both polymer chain diameter and radius of the diffusant (glucose in this case). For hydrogen ions or water, which are much smaller than glucose, the sieving factor is assumed to be negligible, and we simply multiply the respective permeability coefficients, k_H and k_w, by $(1-\phi)$ to account for change in the area available for transport.

Based on these assumptions, the following differential equations govern the hydrogel membrane thickness (L), the flux of hydrogen ions into the membrane, which then become either free intramembrane hydrogen ions (C_H^M) or protons bound to pendant carboxyls (C_{COOH}^M), and the concentration of free hydrogen ions in Cell II (C_H^{II}):

$$\frac{dL}{dt}=k_w(1-\phi)[\ln(1-\phi)+\phi+(\chi_1+\chi_2\phi)\phi^2+\bar{v}_w\rho_o(\frac{\phi_o}{\phi}-\frac{\phi}{2\phi_o})-\bar{v}_wC_s(\lambda+\frac{1}{\lambda}-2)] \quad (1)$$

$$\frac{d}{dt}[L(C^M_{COOH} + C^M_H)] = k_H(1-\phi)[\lambda(C^I_H + C^{II}_H)/2 - C^M_H]$$ (2)

$$\frac{d}{dt}C^{II}_H = \frac{Ak^0_G}{V}e^{-\beta\phi}C_G - \frac{Ak_H}{V}(1-\phi)(\lambda C^{II}_H - C^M_H) - k_{marble}C^{II}_H$$ (3)

In Eqs. (1)-(3), ϕ_0 is the initial polymer volume fraction of a hydrogel membrane (initial thickness L_0, so $L\phi = L_0\phi_0$), C_G is the glucose concentration in Cell I, C_s is the sodium chloride concentration in Cells I and II, V is the volume of Cell II, $\bar{v}_w$ is the molar volume of water, and ρ_0 is the initial concentration of crosslinks in the hydrogel. The RHS of Eq. (2) indicates that the hydrogen ion concentration inside the hydrogel tracks a value that is in "equilibrium" with a hypothetical external medium whose hydrogen ion concentration is the arithmetic mean of the hydrogen ion concentrations in Cell I and Cell II. Since pH=7 in Cell I in the experiments, we set $C^I_H = 10^{-7} M$.

The concentration of carboxylic acid groups in the membrane is determined according to the Langmuir relation,

$$C^M_{COOH} = \frac{\phi}{\phi_0}(\frac{\sigma_0 C^M_H}{K_a + C^M_H})$$ (4)

where σ_0 is the initial concentration of ionizable groups in the membrane (MAA) and K_a is the acid dissociation constant of the pendant carboxyls. The condition for electroneutrality inside the membrane is

$$(1-\phi)(\lambda - 1/\lambda)C_s - (\phi/\phi_0)f\sigma_0 = 0$$ (5)

where the first term corresponds to the mobile charge concentration inside the hydrogel and reflects Donnan partitioning of Na$^+$ and Cl$^-$, and the second term is the fixed charge concentration due to pendant, negatively charged carboxylates. The term f refers to the fraction of MAA groups that are charged, and is given by

$$f = (1 + C^M_H / K_a)^{-1}$$ (6)

Equations (1)-(3) are non-linear ODEs which, when solved along with the algebraic relationships (4)-(6), determine the time evolution of the system. The values of the parameters used for the numerical simulations are, unless otherwise stated: A=3.14cm^2, L_0=250 μm, V=75ml, K_a=10$^{-4.5}$ mol/l, k_G^0=5x10^{-4}cm/sec, k_w =1.5x10^{-2} cm/sec, k_H=5x10^{-3}cm/sec, C_G=50 mM, ρ_0=0.014M, $\bar{v}_w$ =18 ml/mol, β

= 13.5, χ_1 = 0.48, χ_2 = 0.6, and ϕ_0 = 0.3. While some of these values either reflect the experimental conditions or are literature parameters for NIPA/MAA hydrogels (*17*), some parameters were chosen to produce results that are in line with experimental observations. The justification for this parameter selection procedure is that there are known shortcomings in the model which would preclude its ability to make quantitative predictions based on more "accurate" parameter specifications. These shortcomings, along with future remedies, will be discussed below.

Numerical Simulations

The oscillatory behavior of the system is a consequence of the first order phase transition, with hysteresis, in the relationship between the pH in Cell II and the hydrogel membrane's volume. Understanding the pH-swelling behavior of the membrane will provide clues as to the influence of membrane parameters on system dynamics.

Equilibrium swelling behavior of the clamped hydrogel is determined by setting the LHS of Eqs. (1) and (2) to zero, and solving these equations along with Eqs. (5) and (6). Figure 2a shows the one-dimensional swelling behavior of the hydrogel as a function of degree of ionization of the network, $f\sigma_0$. At low ionization, the hydrophobic nature of the hydrogel, as determined by the χ parameter, dominates, and the hydrogel remains in a collapsed, impermeable state. At high values of ionization, the Donnan osmotic pressure is predominant, and the hydrogel is swollen and highly permeable to glucose. A region of bistability exists at intermediate values of ionization. The existence of this region is the basis for membrane hysteresis and oscillatory behavior.

Figure 2b shows equilibrium swelling as a function of pH in Cell II, for various degrees of MAA content (σ_0) in the membrane. At low degrees of MAA substitution, the hydrogel membrane is confined to the branch of the swelling curve corresponding to the collapsed state. At higher MAA contents, the swelling *versus* pH curves develop a sigmoidal shape. With increasing σ_0, the band of bistability shifts in the acidic direction, and the maximal degree of swelling increases. These results are a direct consequence of Eq. (6).

In order to obtain a qualitative understanding of the model's dynamical behavior, Hopf bifurcation plots were generated by linearizing the model equations around steady state, and determining the eigenvalues of the relevant Jacobian (*20*). In these plots, shown below, regions in which steady states are unstable, with the real part of the principal eigenvalue positive, are denoted by "U". Sustained limit cycle behavior is expected in these regions. In regions where the real part of the principal eigenvalue is negative, the system is expected to seek a stable steady state. These regions are marked by "S".

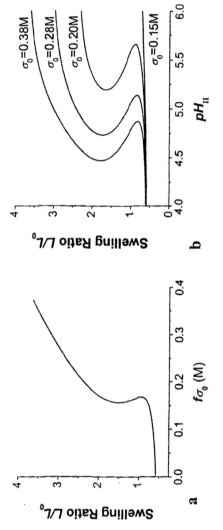

Figure 2. a) Plot of equilibrium swelling of hydrogel, as a function of initial fixed charge density (ionized MAA content) as predicted by model. b) Predicted swelling of the hydrogel as a function of pH in Cell II.

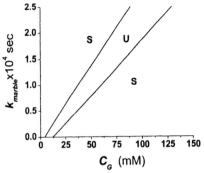

Figure 3. Hopf bifurcation plot for k_{marble} versus C_G, showing regions of local stability (S), and instability leading to oscillatory dynamics (U).

Since MAA doping, glucose concentration in Cell I, and marble reactivity are the most readily controlled quantities in the experimental system, variations in the corresponding parameters, C_G, σ_0, and k_{marble} were studied.

Figure 3 shows regions of oscillatory and non-oscillatory dynamics of the system as a function of k_{marble} and C_G. The region of instability, in which sustained pH oscillations are expected, is a tilted band. For a given glucose concentration in Cell I, C_G, oscillations are predicted over a range of values of k_{marble}. (Experimentally, k_{marble} is controlled by varying the surface area of the marble.) Similarly, for fixed values of k_{marble}, oscillations are expected over a bounded range of glucose concentrations. The banded shape of the region of instability is expected from Eq. (3), where it is seen that increased reactivity of marble must be met with a more rapid production rate of H^+ in Cell II, in order to maintain the steady state pH in Cell II within the region corresponding to bistability of the hydrogel. Increased generation of H^+ in Cell II can only occur by delivering more glucose across the membrane.

Bifurcation analysis was also performed to study the effects of glucose concentration and initial MAA doping (see Figure 4a). A decrease in initial

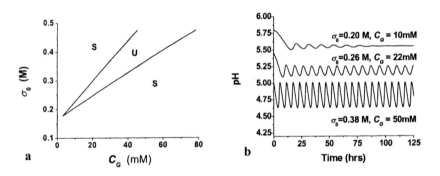

Figure 4. a) Hopf bifurcation plot of initial MAA doping (σ_0) versus glucose concentration (C_G). b) Time profile of pH in Cell II given different values of σ_0 and C_G. $k_{marble}=8x10^{-5} sec^{-1}$.

fixed charge density leads to an increase in the phase transition pH (see Figure 2), and therefore a decrease in the glucose concentration required for oscillation. Figure 4b shows the simulated time profile where the operating pH of the system is increased by reducing σ_0 and C_G. Notice that simulations exhibit both an alkaline shift and a decrease in amplitude of the pH oscillations, with decreasing σ_0. These results are consistent with the experimental results, shown in Figure 1b, pertaining to the effects of decreased MAA doping on pH oscillations.

Discussion

A comprehensive model of the hydrogel/enzyme oscillator will be complex, since it must take into account the presence and transport of several chemical species, and the distributed mechanical response of the hydrogel membrane. In an earlier analysis of the present system (21,22), a highly simplified, electromechanical relay-like model of the hydrogel was considered. This simple, heuristic model, lacked any elements of hydrogel physical chemistry or transport processes between the hydrogel and the reaction compartment (Cell II).

The present model should be regarded as an intermediate step towards a more realistic mathematical representation of the system. We have used an extended Flory-Huggins-Donnan equation of state to model the osmotic pressure inside the hydrogel membrane, and taken the difference in osmotic pressure between the hydrogel and the solutions in Cells I and II to be the driving force for water transport into and out of the membrane. Because pendant carboxylate groups reversibly bind and slow the diffusion of hydrogen ions, we have assumed that this process limits the rate of change of osmotic pressure inside the membrane, and that adjustment of concentrations of other ions in the membrane is essentially instantaneous. Finally, we have used the well-known free volume theory to model the effect of swelling on glucose diffusion in the membrane.

Probably the most severe simplification in the present model is that the membrane is assumed to be homogenous in swelling state and "well stirred" in all concentrations. This simplification leads to a "lumped", ODE-based model. By construction, however, the membrane is always exposed to steep glucose and pH gradients, which should lead to gradients in other species and in the charge and swelling state within the membrane. In particular, glucose transport through the membrane is probably controlled by a skin layer, adjacent to Cell II, which appears and disappears as pH in Cell II oscillates. Intramembrane gradients may also lead to swelling stresses that are not included in the present model of polymer swelling.

Clearly, behavior inside the membrane would be better modeled by partial differential equations, and the next generation of models will be distributed, i.e. based on PDE's. Because of the oversimplifications in the lumped model, it cannot be expected that all parameters will be matched to true physical values, or that all aspects of the data will be explained by this model. Nevertheless, we believe that this intermediate-stage model has heuristic value since it reproduces, at least qualitatively, some of the observed behaviors.

It has been shown that many chemical oscillators involve three elements, namely a constant source of free energy to be dissipated, a bistable element, and a feedback species whose production causes the bistable element to flip between states (14,15). In a typical chemical oscillator, free energy is provided by the inflow of reactants into a continuous stirred tank reactor (CSTR), bistability is provided by an interplay between autoaccelerated production of a reaction intermediate and removal of species by outflow from the CSTR, and feedback is provided by another species in the reaction. The corresponding elements in the hydrogel/enzyme oscillator may be more transparent. Here free energy is dissipated by conversion of glucose to hydrogen ion, which is ultimately removed by release into Cell I and subsequent outflow, bistability is built into the equation of state of the hydrogel membrane, which exhibits hysteresis, and feedback is provided by the hydrogen ions, which alter the charge in the membrane and hence the osmotic swelling force.

Bistability in the model is due to a strong competition between hydrophobic forces, and electrostatic/Donnan forces. The hydrophobic force is cooperative in this model; cooperativity is manifested in the increase in the interaction parameter, χ, with polymer volume fraction.

In the present study, we considered the effect of degree of doping of the membrane with MAA, which is represented by the parameter σ_0. From Figure 2, it is evident that when σ_0 is below a critical value, the membrane will remain in its collapsed state, because membrane charge cannot overcome membrane hydrophobicity.

With increased MAA substitution, the relation between $f\sigma_0$ and membrane swelling moves into the bistable region (see Figure 2), and oscillations are now possible. As σ_0 is increased further, the value of f needed to keep the system in or near the bistable region decreases, and from Eq. (6) this shifts the bistable region in the acidic direction, as illustrated in Figure 2b. Consequently, an increase in MAA doping promotes an acidic shift in the range of pH oscillations (Figures 1b and 4b). Finally, such an acidic shift with increasing σ_0 requires that more glucose be delivered into Cell II. This observation is made quantitative, for the present model, by the bifurcation diagram in Figure 4a.

It may be noticed that even though the pH *versus* swelling characteristic for σ_0=0.2 M exhibits hysteresis (Figure 2b), simulated pH oscillations decay (Figure 4b). We note however that for σ_0=0.2 M, the range of glucose concentrations compatible with oscillations becomes very narrow (Figure 4a), and the combination σ_0=0.2, C_G=10 mM lies just outside that range, thus explaining the decay to steady state in the simulation. (Lowering C_G to 8 mM does produce sustained but weak pH oscillations, not shown.)

References

1. Flory, P. J. *Principles of Polymer Chemistry*; Cornell University Press: Ithaca, NY, 1953.
2. Shibayama, M.; Tanaka, T. *Adv. Polym. Sci.* **1993**, *109*, 1.
3. Siegel, R. A. pH-Sensitive Gels: Swelling Equilibria, Kinetics, and Application for Drug Delivery. In *Pulsed and Self-Regulated Drug Delivery*; Kost, J., Ed.; CRC Press: Boca Raton, FL, 1990.
4. Firestone, B. A.; Siegel, R. A. *J. Biomat. Sci. Polym. Ed.* **1994**, *5*.
5. Ishihara, K.; Kobayashi, M.; Ishimaru, N.; Shinohara, I. *Polym. J.* **1984**, *16*, 625.
6. Kataoka, K.; Miyazaki, H.; Bunya, M.; Okano, T.; Sakurai, Y. *J. Amer. Chem. Soc.* **1998**, *120*, 12694.
7. Sato-Matsuo, E.; Tanaka, T. *J. Chem. Phys.* **1988**, *89*, 1695.
8. Baker, J. P.; Siegel, R. A. *Macromol. Rapid. Commun.* **1996**, *17*, 409.
9. Yoshida, R.; Ichijo, H.; Hakuta, T.; Yamaguchi, T. *Macromol. Rapid Commun.* **1995**, *16*, 305.
10. Yoshida, R.; Tanaka, M.; Onodera, S.; Yamaguchi, T.; Kokufuta, E. *J. Phys. Chem. A.* **2000**, *104*, 7549.
11. Misra, G. P.; Siegel, R. A. *J. Controlled Release* **2002**, *81*, 1.
12. Dhanarajan, A. P.; Misra, G. P.; Siegel, R. A. *J. Phys. Chem.* **2002**, *106*, 8835.
13. Sekimoto, K. *Phys. Rev. Lett.* **1993**, *70*, 4154.
14. Epstein, I. R.; Showalter, K. *J. Phys. Chem.* **1996**, *100*, 13132.
15. Hunt, K. L. C.; Hunt, P. M.; Ross, J. *Annu. Rev. Phys. Chem.* **1990**, *41*, 409.
16. Rábai, G.; Hanazaki, I. *J. Phys. Chem.* **1996**, *100*, 10615.
17. Hirotsu, S. *Phase Transitions* **1994**, *47*, 183.
18. Grimshaw, P. E.; Nussbaum, J. H.; Grodzinsky, A. J.; Yarmush, D. M. *J. Chem. Phys.* **1990**, *93*, 4462.
19. Yasuda, H.; Lamaze, C. E.; Peterlin, A. *J. Poly. Sci. Pt. A-2* **1971**, *9*, 1117.
20. Gray, P.; Scott, S. K. *Chemical Oscillations and Instabilities*; Clarendon: Oxford, 1990.
21. Zou, X.; Siegel, R. A. *J. Chem. Phys.* **1999**, *110*, 2267.
22. Leroux, J.-C.; Siegel, R. A. *Chaos* **1999**, *9*, 267.

Chapter 5

A Model for Self-Oscillating Miniaturized Gels

P. Borckmans, K. Benyaich, A. De Wit, and G. Dewel

Service de Chimie-Physique, CP231, Université Libre de Bruxelles,
Bruxelles B–1050, Belgium

In many works on chemical patterns gels inactive regarding the chemistry involved have been used as reaction medium to suppress hydrodynamic convection while allowing the control of the system far from equilibrium. Instead, in this study, we consider the case of an active gel the behavior of which is influenced by some products of the reaction. We propose a simple model where the gel exhibits mechanical oscillations in response to a chemical reaction.

Introduction

Gels are cross-linked networks of polymers immersed in a fluid medium. It is now well known that they can exhibit large volume changes in response to many different stimuli: temperature, solvent composition, pH, electric fields. The universality of this volume phase transition of gels has now been clearly established. These stimuli-responsive gels have opened a new field of research by generating numerous experimental and theoretical works; they also pave the way for a variety of new technologies (*1*).

Since they are open systems that can exchange chemical species with their surrounding solvent, gels can also play the role of chemical reactors. In this framework, the design of open spatial gel reactors has allowed well controlled experimental studies of chemical patterns such as chemical waves or Turing structures (*2*). They are made of a thin film of gel in contact with one or two continuous stirred tank reactors that sustain controlled nonequilibrium conditions.

In these experiments, the volume of the confined gel is constant; its main role is to damp hydrodynamical motions that would otherwise perturb the chemical intrinsic patterns. More recently it has been shown experimentally that the coupling of a volume phase transition with a chemical oscillator can generate a self-oscillating gel (3, 4). More precisely, if one of the chemical species taking part in the chemical reaction modifies the threshold for the phase transition, then the time periodic variation of this concentration can generate autonomous swelling-deswelling cycles of the gel even in absence of any external stimuli (5, 6). This device thus provides a novel biomimetic material with potential biomedical and technical applications.

The main purpose of this paper is to present a simple theoretical model to describe this interesting phenomenon which provides a further example of the synergy between an equilibrium phase transition and a pattern forming instability. It is based on a Landau type equation for the polymer volume fraction coupled to a simple two variables reactive system that, on its own, can undergo a Hopf bifurcation giving rise to chemical oscillations of the limit cycle type. These oscillating systems have now been extensively studied both from the theoretical and experimental points of view (7).

For the sake of simplicity, we here study gels the dimensions of which are smaller than the characteristic wavelength of the chemo-elastic waves that can also appear in large systems. This enables us to consider homogeneous and uniform systems in agreement with the experiments performed recently on miniaturized oscillating gels.

We have first verified that, in absence of chemical reactions, our dynamical equation for the gel reproduces the dramatic slowing downs in the transition rates which have been observed near the critical point or the spinodal limits of hydrogels. In the last section, we discuss an example of the instabilities that give rise to the self-oscillating behavior of the coupled gel-chemical system and we exhibit corresponding bifurcation diagrams.

Volume phase transition of gels

We first outline the thermodynamic theory of stimuli responsive gels (8). The free energy is given as a sum of contributions due to the mixing of the solvent and polymer matrix, its elasticity and the presence of counter ions

$$\Delta F = \Delta F_{mix} + \Delta F_{el} + \Delta F_{ion} \tag{1}$$

In terms of the polymer volume fraction ϕ, the Flory-Huggins theory (9) gives

$$\Delta F_{mix} = \frac{kT}{v_1} V\left[(1-\phi))\ln(1-\phi) + \chi\phi(1-\phi)\right] \tag{2}$$

where V and v_1 are respectively the volume of the gel and the molar volume of the solvent. The polymer-solvent Flory interaction parameter χ is a function not only of the temperature but it may also depend on the concentrations of some soluted chemical species (10). On the basis of the simple rubber elasticity theory ΔF_{el} can be expressed as

$$\Delta F_{el} = \frac{3kTV_0v_0}{2}\left[(\frac{\phi_0}{\phi})^{2/3} - 1 - \frac{2}{3}B\ln(\frac{\phi_0}{\phi})\right] \tag{3}$$

where V_0 and ϕ_0 are the volume and polymer volume fraction in a reference state ($\phi/\phi_0 = V_0/V$ $\phi/\phi_0 = V_0/V$) and v_0 the cross link number density. B is a debated coefficient which we set to unity according to Flory's treatment.
Finally, ΔF_{ion} includes the translational entropy of the counter ions of density v_i

$$\Delta F_{ion} = -kTV_0v_i\ln(\frac{\phi_0}{\phi}) \tag{4}$$

The osmotic pressure

$$\Pi = -(\frac{\partial F}{\partial V})_T = \frac{\phi}{V}(\frac{\partial F}{\partial \phi})_T \tag{5}$$

plays an important role in volume phase transitions. In the model, it is given by the following explicit expression

$$\Pi = -\frac{kT}{v_1}\left[\phi + \ln(1 - \phi) + \chi\phi^2\right] + kTv_i(\frac{\phi}{\phi_0}) + kTv_0\left[\frac{1}{2}(\frac{\phi}{\phi_0}) - (\frac{\phi}{\phi_0})^{1/3}\right] \tag{6}$$

For small values of ϕ, the elastic contribution serves to limit the degree of swelling. The osmotic pressure must be zero for the gel at equilibrium with the surrounding solvent. As shown in Figure 1, by varying χ, the gel can undergo a phase transition from swollen to collapsed state (and vice-versa). In such process the system exhibits an asymmetric hysteresis loop. In the case of gels, due to the long range behavior of the elastic energy of deformation (proportional to the volume of the system) the mean-field Van der Waals type theory applies and the transition does not take place at the Maxwell point $\chi = \chi_m$ where the free energy of the swollen branch becomes equal to that of the collapsed state. It rather occurs at (or near) the spinodal line (11, 12). This leads to observable hysteresis effects in the transition. Indeed experiments have reported differences of temperatures of up to 10 degrees between the temperature at which the collapse occurs and that where swelling takes place. Such hysteresis are also obtained when varying the pH (13, 14). Thermodynamic stability implies that the elastic bulk modulus $K = \phi(\partial\Pi/\partial\phi)_T \geq 0$. The condition $K = 0$ determines the spinodal points χ_{sc} and χ_{ss} shown in Figure 1. They respectively correspond to the marginal stability points of the collapsed and the swollen state. The locus of these points when the temperature or the solvent composition is varied determines the spinodal curve. The maximum of this curve corresponds to a critical point at which the volume phase transition becomes continuous. When the corresponding osmotic pressure is equal to zero, the system exhibits a critical endpoint at which the first three derivatives of the free energy with respect to V must vanish. Tanaka found such a critical endpoint in a polyacrylamide gel (15). He succeeded in reducing the volume discontinuity at zero osmotic pressure by varying the composition of the solvent.

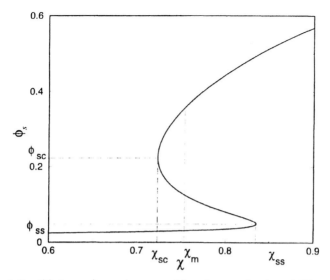

Figure 1. Equilibrium volume phase transition diagram for the gel. The polymer volume fraction ϕ is represented as a function of Flory's interaction parameter χ.

These conditions allow to determine the critical values of ϕ_c, γ_c and $\chi_c(T_c)$. As they are not required in the following we will not give here their values in terms of the parameters defining the free energy. Keeping $\gamma = \gamma_c$, near the critical point Π assumes the following standard form $\Pi = a_0\left(\chi_c - \chi\right) + b_0\left(\phi - \phi_c\right)^3$, where a_0 and b_0 are constants independent of ϕ and χ. One can then define a critical index δ such that $\left(\phi - \phi_c\right) \propto \left|\chi - \chi_c\right|^\delta$ with $\delta = 1/3$. Note that here, $\chi - \chi_c$, plays a role analogue to that of the magnetic field in spin systems. The bulk modulus tends to zero as $\chi \to \chi_c$ according to $K \propto \left(\phi - \phi_c\right)^2 \propto \left|\chi - \chi_c\right|^{2/3}$. Similarly near the spinodal points χ_{sc} and χ_{ss}, the osmotic pressure takes the characteristic expression of a saddle-node bifurcation $\Pi = a_{si}\Delta\chi + b_{si}\left(\phi - \phi_{si}\right)^2$ where $\Delta\chi = \left|\chi - \chi_{si}\right|$ with $i \equiv c$ or s and the bulk modulus now becomes $K = \left|\Delta\chi\right|^{1/2}$.

In general the polymer-solvent interaction depends upon the polymer volume fraction. This is taken into account by a power series expansion $\chi = \chi_1 + \chi_2\phi + \chi_3\phi^2 + \ldots$ (*16*). It has therefore been shown that a strong dependence on ϕ (sufficiently large value of χ_2) can induce a discontinuous transition even in the case of nonionic gels (*17*).

Relaxation kinetics

The polymer volume fraction ϕ plays the role of a nonconserved order parameter. In (small) homogeneous and isotropic gels, its time evolution is assumed to satisfy the following Landau type equation

$$\frac{d\phi}{dt} = -\Gamma\frac{\delta\Delta F}{\delta\phi} = -\Gamma\frac{V_0\phi_0}{\phi^2}\Pi \tag{7}$$

where the kinetic coefficient Γ fixes the time scale. In this description, it is the osmotic pressure that provides the thermodynamic force for the volume phase transition. Obviously, the stationary solutions of Eq. (7) correspond to the thermodynamic states $\Pi(\phi_s) = 0$. Within the above model

$$\frac{d\phi}{dt} = \frac{\alpha}{\phi^2}\left[(1-\gamma)\phi + \ln(1-\phi) + \chi\phi^2 + \beta\phi^{1/3}\right] \tag{8}$$

where $\alpha = kTV_0\phi_0/v_1$, $\beta = v_0v_1/\phi_0^{1/3}$ and $\gamma = \left(\frac{v_0}{2} + v_i\right)v_1/\phi_0$.

For numerical computations we use typical values found in the literature for gels considered in the study of volume phase transitions.

The theory based on Eq. (8) differs from the standard kinetic description of gels volume phase transition which is based on a diffusion-type equation for the displacement vector $u(r,t)$ of a gel element (18). There that linear equation is derived from the theory of elasticity of the gel. It can only describe the relaxation of small amplitude long wavelength inhomogeneities around a homogeneous state (swollen or shrunken) but it cannot account for the large volume changes that take place in some hydrogels. It does not contain an intrinsic driving force for the volume changes but this must be introduced through the initial condition which is thus different in the cases of swelling and shrinking. They are therefore appropriate to describe the response of the gel to a stimulus such as a change in solvent composition or temperature but they cannot describe swelling-deswelling cycles that have been observed in self-oscillating systems. Finally, as pointed out by Onuki (19), that diffusion equation overlooks the slow mode representing homogeneous swelling or shrinking with minor density inhomogeneities. It has also been criticized on other grounds (12, 20).

Relaxation methods provide a useful tool to probe the dynamical properties of physico-chemical systems. In this framework, the knowledge concerning how fast a gel shrinks or swells is essential for many technological applications. The relaxation time τ of small perturbations about a steady state ϕ_s is obtained by linearizing Eq. (8) to give

$$\tau = \frac{\phi_s^2}{\Gamma\phi_0 V_0}\left(\frac{\partial\Pi}{\partial\phi}\bigg|_{\phi_s}\right)^{-1} = \frac{\phi_s^3}{\Gamma\phi_0 V_0 K} \tag{9}$$

The present dynamical description is thus compatible with the thermodynamic stability condition $K>0$.

From Eq. (9) we see that relaxation processes present a dramatic slowing down in the vicinity of instabilities such as critical or cloud points where $K\to0$. The relaxation time indeed becomes infinitely large when the critical endpoint is approached both from above and from below. From Eq. (9) one can define a dynamical critical index $\tau \propto |\chi - \chi_c|^{-n_c}$ with $n_c = 2/3$. Similarly, the transition rate (i.e. the inverse of the relaxation time) decreases when the final state to which the system is pumped comes close to the hysteresis limits (χ_{si}, ϕ_{si} with i $\equiv$

c or s). This behavior $\tau \propto |\chi - \chi_{si}|^{-n_s}$ can now be characterized by a spinodal index $n_s = 1/2$, according to Eq. (9). These phenomena of critical and spinodal slowing down have been experimentally observed on sub millimeter spherical NIPA gels (*21, 22*).

Finally, gels can also exhibit a slow relaxation when they are driven slightly outside the bistability region. More precisely, if the system is initially prepared in a state such that $\phi_0 > \phi_{si}$ and $\chi < \chi_{sc}$, it undergoes a slowing down during its relaxation towards the swollen state when it is brought closer to the hysteresis limit (χ_{sc}, ϕ_{sc}). On this plateau (Figure 2), the system presents a slow power law decrease proportional to $(\chi_{sc} - \chi)\tau_{pl}$. The slope and the lifetime of this plateau are very sensitive to the proximity of the hysteresis limit χ_{sc}. The lifetime is given by $\tau_{pl} \propto \kappa(\chi - \chi_{sc})^{-1/2}$ where κ is a measure of the curvature of the hysteresis loop at the limit point. This dynamical behavior has also been experimentally observed during the swelling process of NIPA gels (*22*). After a

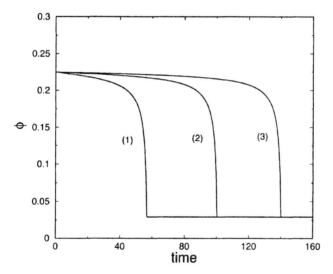

Figure 2. Relaxation curves slightly beyond the hysteresis limit χ_{sc} for the same initial condition but for decreasing distance from the spinodal point. $\alpha=1$, $\beta=0.003$, $\gamma=0.038$, $\chi_{sc}=0.7233$. (1)$\chi= 0.723$ (2)$\chi=0.7232$ (3)$\chi=0.72322$.

temperature jump, the gel starts at first to swell slowly until the process speeds up in the final stages. Both the inverse of the slope of the plateau and its lifetime are shown to diverge at the threshold. This mechanism can in principle show up both in the swelling and the shrinking processes. However as shown in Figure 1, the hysteresis loop in the gel systems is highly asymmetric. The curvature at the marginal stability limit of the collapsed state is smaller than at the other limit point. As a result at the same distance from the limit points the plateau lifetime is much larger and can thus more readily be observed in the swelling process.

This phenomenon of spinodal slowing down presents generic features that only depend on the nature (saddle-node) bifurcation that delimit the domain of bistability and has for instance also been observed in chemical systems (*23*). It would thus be interesting to experimentally test the various scalings derived in this section.

Self-oscillating gels

We now discuss the coupling between the swelling-deswelling dynamics as described above and a potentially oscillating chemical reaction. As chemical process we could have chosen realistic oscillating reactions that abound in the specialized literature such as, for instance, the Belousov-Zhabotinsky reaction or the bromate/sulfite/ferrocyanide pH oscillator (7). Kinetic schemes thereof may in some instances lead to semi-quantitative descriptions of the oscillations without resorting to the introduction of the ionic character of these reactions that occur in solutions. Nevertheless in this first approach and because the couplings with the gel have not yet been characterized we have chosen the formal Brusselator model that is known to exhibit autonomous oscillations that are well documented (24). For the ease of presentation, we therefore consider a neutral gel for which the discontinuous volume phase transition is induced by the dependence of the Flory interaction parameter on polymer volume fraction. We also suppose that some species involved in the reaction can influence the volume phase transition through the dependence of the expansion coefficients $\chi_1, \chi_2, \chi_3...$ on their concentrations. There lies the main coupling between the two subsystems. The governing kinetic equations then take the following form

$$\frac{d\phi}{dt} = -\Gamma \frac{\delta \Delta F(X,Y,\phi)}{\delta \phi}$$

$$\frac{dX}{dt} = f(X,Y) - \frac{X}{V}\frac{dV}{dt} \tag{10}$$

$$\frac{dY}{dt} = g(X,Y) - \frac{Y}{V}\frac{dV}{dt}$$

The model we explicitly consider is thus

$$\frac{d\phi}{dt} = \frac{\alpha}{\phi^2}\left[(1-\gamma)\phi + \ln(1-\phi) + \left(\chi_1^0 + \gamma_1 Y + \chi_2(2\phi-1)\right)\phi^2 + \beta\phi^{1/3}\right]$$

$$\frac{dX}{dt} = A - (B+1)X + X^2Y + \frac{X}{\phi}\frac{d\phi}{dt} \tag{11}$$

$$\frac{dY}{dt} = BX - X^2Y + \frac{Y}{\phi}\frac{d\phi}{dt}$$

where the parameters α, β and γ have been defined earlier; γ_1 describes the influence on the gel of Y, the sole species of the chemical reaction which changes the solvent properties through $\chi_1 = \chi_1^0 + \gamma_1 Y$; A and B are chemical control parameters. In absence of phase transition chemical oscillations occur through a Hopf bifurcation at $B_H = 1 + A^2$. The last terms in the chemical kinetic equations are the concentration-dilution contributions taking into account the volume variations of the "gel reactor". The existence of these contributions imply that the role of chemistry is not, in general, merely that of a parametric forcing of the gel dynamics. We have a new dynamical system at hand.

Parameter space is thus multidimensional so that we only discuss some illustrative cases. For a first set of values of the parameters Figure 3 describes the

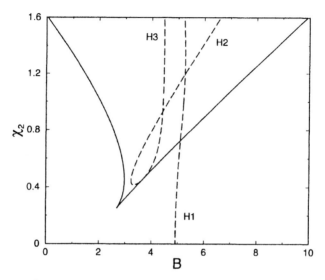

Figure 3. Bifurcation lines in the (χ_2, B) plane of parameter space for $\alpha=0.2$, $\beta=0.00758$, $\gamma=0.00213$, $\chi^0{}_1=0.3$, $\gamma_2=0.32$, $A=2$. Plain lines are the locus of saddle-node bifurcations whereas the dotted lines are loci of Hopf bifurcations.

bifurcations in the (χ_2, B) plane. Let us recall that χ_2 controls the size of the discontinuity of the volume phase transition in the absence of a chemical process while B is the standard bifurcation parameter of the Brusselator.

When χ_2 is sufficiently large, the swollen and collapsed states coexist for a range of values of B inside the cusped region. As a result of the coupling, each state can undergo a Hopf bifurcation leading to mechano-chemical oscillations. The loci of the bifurcations leading to these oscillations are represented by the foliated curve H_2H_3 and the line H_1 that remains near the value of B_H for the chosen values of the parameters. For these values, when $\chi_2 = 1.02$, the states of the polymer volume fraction are shown in Figure 4. For small values of B the swollen state (small value of ϕ_s) is stable and may eventually coexist with the shrunken state. However at H_2 the system undergoes a supercritical Hopf bifurcation to temporal oscillations. Meanwhile the collapsed state also undergoes a supercritical bifurcation at H_1. The oscillations of the swollen state do not however persist as their stable limit cycle annihilates with an unstable limit cycle emanating from H_3. For larger values of B only the oscillations of the shrunken state persist. The oscillations are of small amplitude as shown on Figure 5. There for small times we show the oscillations of the polymer volume fraction in the swollen state for $B=6.5$. After 20 units of time B is stepped up to 8 (beyond the collision of the stable and unstable limit cycles). The polymer

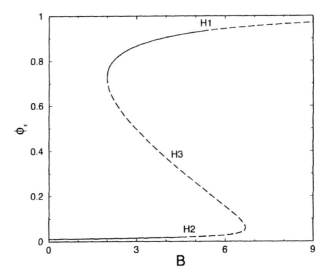

Figure 4. Polymer volume fraction ϕ_s as a function of B. Solid and dotted lines respectively represent stable and unstable steady states. The parameters are as in Figure 3 and $\chi_2=1.02$.

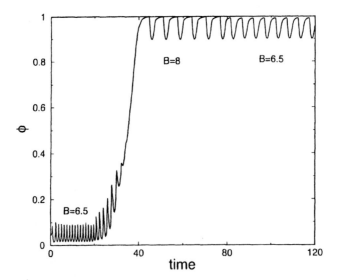

Figure 5. Mechano-chemical oscillations and birhythmicity. Parameters as in Figure 3.

therefore shrinks as its volume fraction transits to oscillations around the upper branch. When the system has settled in this new state, at t = 80, B is then stepped down to 6.5 again. The polymer volume fraction keeps on oscillating around its shrunken state albeit with a smaller amplitude as one comes nearer the H_1 Hopf bifurcation. This also shows that the system exhibits birhythmicity as, for instance at $B = 6.5$, oscillations around the swollen and collapsed state coexist.

Although it is difficult to be quantitative at this stage the smallness of the oscillations can be tracked to the relation between the characteristic time of the gel, related to α and that of the chemistry. A measure of the last one is the inverse of the critical frequency of the limit cycle that is equal to A for the Brusselator. If this is so one may intuitively argue that when α is too small or $1/A$ too large, chemistry "recalls" the gel before it has swollen or shrunken as much as it could. To test this we have measured the amplitude of the oscillations for the same conditions as before changing only the value of α. The result is shown on Figure 6. Indeed the amplitude of the oscillations increase.

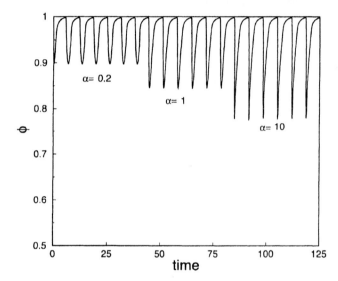

Figure 6. Influence of α on the amplitude of the oscillations of ϕ for $B=8$.
Parameters as in Figure 3.

To show the reverse effect of varying A we have chosen somewhat different conditions where the critical point is embedded in the folium and the oscillations of the swollen state are more of a relaxation type. This is shown in Figure 7. The amplitudes of the oscillations increase as A decreases. It therefore seems that the product of the gel relaxation time τ by the characteristic frequency ω of the chemical oscillations, $P = \omega\tau$, provides the crucial parameter that determines the amplitude of these oscillations.

As illustrated in Figures 5 and 7 the amplitude and frequencies of the gel oscillations depend on the values of the chemical parameters. Figure 8 also shows that there is no phase shift in the oscillations of the chemical concentration and the polymer volume fraction. These effects has been observed in the recent experiments (*4, 6, 25*).

The fact that we are dealing with a new dynamical system is asserted by the fact that it may oscillate for reactant concentrations that are such that the chemical system on its own does not exhibit oscillations as it lies below its Hopf bifurcation limit.

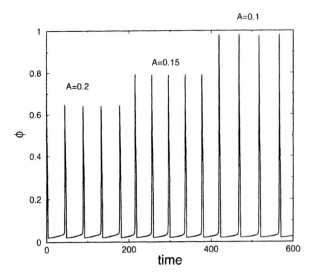

Figure 7. Influence of A on the amplitude of the oscillations of φ for B=0.026.
$\alpha=0.7$, $\beta=0.00758$, $\gamma=0.00213$, $\chi^0_1=0.5$, $\gamma_2=0.2$.

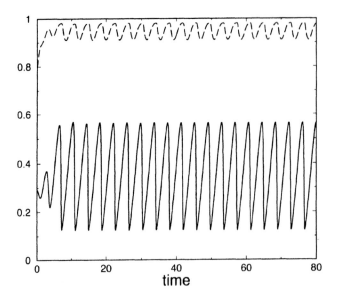

Figure 8. In phase periodic oscillations of the polymer volume fraction around its shrunken state(dotted line) and the concentration of chemical species Y (plain line that represents Y/10).

Conclusions

We have presented a very simple model that hints at how an oscillating chemical reaction can be used to drive the periodic oscillation of a piece of gel and thereby transduce chemical energy into mechanical work.

Our toy model however does not take into account the all important spatial effects originating from the diffusive properties of the matrix of the gel (and related to the elastic properties) and those of the reactive solute species. These will introduce further characteristic time scales and new length scales in the problem. The inclusion of such terms should allow for the description of the chemical waves observed by Yoshida et al (25).

In the absence of chemistry various approaches have been proposed to take such spatial effects into account (12, 18, 26, 27, 28, 29, 30, 31), but their relation to one another has not been tested thoroughly nor to what extend they include the spatial phenomena related to the volume phase transition as studied separately (19, 32). Along those lines a model for a sphere of gel presenting small temporal oscillations provoked by a stationary chemical reaction is presented in this volume with related experimental results (33).

The determination of the precise nature of the couplings between the reacting chemicals and the gel matrix, as well as their intensities should present interesting experimental and theoretical challenges.

Acknowledgement. P.B., G.D. and A.D. received support from the FNRS (Belgium). We also thank the Fondation Universitaire Van Buuren and the Tournesol Program (CGRI) for financial support. K.B. also received a de Meurs-François grant. We benefited from numerous discussions with J. Boissonade, E. Dulos, F. Gauffre and P. De Kepper and are grateful for Professor I. Prigogine's encouragements.

References

1. *Responsive Gels: Volume Transitions*, Editor, Dusek, K.; Adv. Polymer Sci. 109 & 110; Springer: Berlin, **1993**, pp1-275 & 1-269.
2. De Kepper, P.; Dulos, E.; Boissonade, J.; De Wit, A.; Dewel, G.; Borckmans, P. *J. Stat. Phys.* **2000**, *101*, 495-508.
3. Yoshida, R.; Ichijo, H.; Hakuta, T.; Yamaguchi, T. *Macromol. Rapid Commun.* **1995**, *16*, 305-310.
4. Crook, C.J.; Smith, A.; Jones, R.A.L. Jones, Ryan, A. J. *Phys. Chem. Chem. Phys.* **2002**, *4*, 1367-1369.
5. Yoshida, R.; Takahashi, T.; Yamagushi, T.; Ichijo, H. *J. Am. Chem. Soc.* **1996**, *118*, 5134-5135.
6. Yoshida, R.; Kokufuta, E.; Yamagushi, T. *Chaos* **1999**, *9*, 260-266.
7. Epstein, I.; Pojman, J.A. *An Introduction to Nonlinear Chemical Dynamics*; Oxford University Press: Oxford, UK, 1998; pp1-392.
8. Shibayama, M.; Tanaka, T. In *Responsive Gels: Volume Transitions*, Editor, Dusek, K.; Adv. Polymer Sci. 109; Springer: Berlin, **1993**, pp 1-62.
9. Flory, P. *Principles of Polymer Chemistry*; Cornell University Press: Ithaca, NY, 1953.
10. Hirotsu, S. *J. Chem. Phys.* **1988**, *88*, 427-431.

70

11. Hochberg, A.; Tanaka, T.; Nicoli. D. *Phys. Rev. Lett.* **1979**, *43*, 217-219.
12. Tomari, T.; Doi, M. *Macromolecules* **1995**, *28*, 8334-8343.
13. Siegel, R.A. In *Responsive Gels: Volume Transitions*, Editor, Dusek, K.; Adv. Polymer Sci. 109; Springer: Berlin, 1993, pp233-267.
14. Suzuki, A.; Suzuki, H. *J. Chem. Phys.* **1995**, *103*, 4706-4710.
15. Tanaka, T. *Phys. Rev. Lett.* **1978**, *40*, 820-823.
16. Erman, B.; Flory, P.J. *Macromolecules*, **1986**, *19*, 2342-2353.
17. Hirokawa, Y.; Tanaka, T. *J. Chem. Phys.* **1984**, *81*, 6379-6380.
18. Tanaka, T.; Fillmore, D.J. *J. Chem. Phys.* **1979**, *70*, 1214-1218.
19. Onuki, A. In *Responsive Gels: Volume Transitions*, Editor, Dusek, K.; Adv. Polymer Sci. 109; Springer: Berlin, 1993, pp 63-121.
20. Komori, T.; Sakamoto, R. *Colloid Polym. Sci.* **1989**, *267*, 179-183.
21. Tanaka, T.; Sato, E.; Hirokawa, Y.; Hirotsu, S.; Peetermans, J. *Phys. Rev. Lett.* **1985**, *55*, 2455-2458.
22. Matsuo, E.S.; Tanaka, T. *J. Chem. Phys.* **1988**, *89*, 1695-1703.
23. Dewel, G.; Borckmans, P. In *Spatial Inhomogeneities and Transient Behaviour in Chemical Kinetics*, Editors, Gray, P.; Nicolis, G.; Baras, F.; Borckmans, P.; Scott, S.K.; Proceedings in Nonlinear Science; Manchester University Press: Manchester, UK, **1990**, pp205-222, and references therein.
24. Nicolis, G.; Prigogine, I. *Self-organization in non-equilibrium systems*; Wiley: New York, 1977, pp1-461.
25. Miyakawa, K.; Sakamoto, F.; Yoshida, R.; Kokufuta, E.; Yamaguchi, T. *Phys. Rev. E* **2000**, *62*, 793-798.
26. Johnson, D.L. *J. Chem. Phys.* **1982**, *77*, 1531-1539.
27. Sekimoto, K. *J. Phys. II* **1991**, *1*, 19-36.
28. Bruinsma, R.; Rabin, Y. *Phys.Rev. E* **1994**, *49*, 554-569.
29. English, A.E.; Tanaka, T.; Edelman, E.R. *J. Chem. Phys.* **1997**, *107*, 1645-1654.
30. Bisschops, M.A.T.; Luyben, K.Ch.A.M.; van der Wielen, L.A.M. *Ind. Eng. Chem. Res.* **1998**, *37*, 3312-3322.
31. Achilleos, E.C.; Christodoulou, K.N.; Kevrekidis, I.G. *Computational & Theoretical Polymer Science* **2001**, *11*, 63-80.
32. Onuki, A.; Puri, S. *Phys. Rev. E* **1999**, *59*, R1331-1334.
33. Gauffre, F.; Labrot, V.; Boissonade, J.; De Kepper, P. (this volume).

Chapter 6

Measurement of Force Produced by a pH-Responsive Hydrogel in a pH Oscillator

Colin J. Crook[1,*], Ashley Cadby[1], Andrew Smith[1],
Richard A. L. Jones[2], and Anthony J. Ryan[1,*]

Departments of [1]Chemistry and [2]Physics, University of Sheffield,
Sheffield S3 7RH, United Kingdom

A Landolt pH-oscillator based on a bromate/sulfite/ferrocyanide reaction has been developed with a room temperature period of 20 minutes and a range of $3.1 \leq pH \leq 7.0$. This system has been used to drive periodic oscillations in volume in a pH responsive hydrogel. A continuously stirred, constant volume, tank reactor was set-up in conjuction with a modified JKR experiment and is used to show that the combination of a pH oscillator and a pH responsive hydrogel can be used to generate measurable force.

Introduction

For many years, the ability of some polymers to change conformation in response to an external stimuli has been a source of great interest from both experimental and theoretical standpoints (*1*). These polymers, whin the form of brushes and gels have the ability to change their shape, making them candidates for a new generation of intelligent materials with sensor, processor and actuator functions. A wide variety of different stimuli-responsive gels have been developed for specific applications such as drug release and actuators for artificial muscle (*2*). Microscopic devices have been built and pH sensitive gels have been used to build "flow-sorter" valves for fluidic systems (3) where the volume transition of the gels open and close 200 μm channels.

Studies carried out by Yoshida and coworkers have coupled this phenomena with oscillating chemical reactions (such as the Belousov-Zhabotinsky, BZ, reaction) to create conditions where "pseudo" non-equilibrium systems which maintain rhythmical oscillations can be demonstrated, in both quiescent (4) and continuously stirred reactors (5). The ruthenium complex of the BZ reaction was introduced as a functional group into poly(N-isopropyl acrylamide), which is a temperature-sensitive polymer. The ruthenium group plays it's part in the BZ reaction, and the oxidation state of the catalyst changes the collapse temperature of the gel. The result is, at intermediate temperature, a gel whose shape oscillated (by a factor of 2 in volume) in a BZ reaction, providing an elegant demonstration of oscillation in a polymer gel. This system, however, is limited by the concentration of the catalyst which has to remain relatively small, and hence the volume change is small.

We present a system which consists of a polyelectrolyte material and a Landolt reaction that oscillates pH (6). Poly (methacrylate acid), a weak polyacid, will aquire a net charge at low pH owing to dissociation of the acid groups, and hence the charged chains repulse each other causing them to stretch away from each other. Thus by oscillating the pH one can induce a large macroscopic oscillation in the dimensions of the gel. We have recently reported our monitoring of gel swelling by optical microscopy in this system (7) and have observed that the volume changes by a factor of at least 6.

Of course, in order to be used as synthetic 'muscle' and hence to fabricate practical devices, it is necessary to show that these systems are capable of producing measurable force. In the first study of its kind, this is demonstrated by the use of a relatively simple apparatus.

The Bromate/Sulfite/Ferrocyanide pH Oscillator

The work outlined in this chapter makes use of pH as the stimuli for the system. This stimuli is delivered by a Bromate/Sulfite/Ferrocyanide pH oscillator developed by Edblom, Orban and Epstein (8). The benefits of the system are as follows;

1. Reliable oscillation at room temperature.
2. Large pH range.
3. Stays at pH extremes for approximately equal time to allow gel to change conformation.

The mechanism of the process is very complicated, consisting of many mechanistic steps, but may be viewed at a simplified level as consisting of two subsystems; the oxidation of sulfite by bromate (1) and the oxidation of ferrocyanide by bromate (2).

$$BrO_3^- + 3HSO_3^- + H^+ \longrightarrow Br^- + 3SO_4^{2-} + 4H^+ \tag{1}$$

$$BrO_3^- + 6Fe(CN)_6^{4-} + 6H^+ \longrightarrow Br^- + 6Fe(CN)_6^{3-} + 3H_2O \tag{2}$$

The first process serves as the proton production mechanism (positive feedback) with an appropriate "induction time", while the second acts as the negative feedback reaction consuming protons.

The reactions were conducted in an open continuously stirrer tank reactor (CSTR) at room temperature, with a peristaltic pump supplying the feed solutions of potassium bromate, sodium sulfite, potassium ferrocyanide and sulphuric acid; and also pumping out bulk solution to keep the volume in the CSTR constant. Measurements of pH were made using a standard gel-filled probe, and the output from the meter monitored by a PC. A typical pH profile is shown in Figure 1.

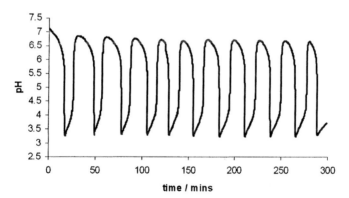

Figure 1. The output of a pH meter immersed in a CSTR with the Landolt oscillator. Input concentrations: [SO₃²⁻] = 0.075 M, [BrO₃⁻] = 0.065 M, [Fe(CN)₆⁴⁻] = 0.02 M, [H₂SO₄] = 0.01 M. Residence time = 1000s, at room temperature. (Reproduced from reference 7 by permission of The Royal Society of Chemistry on behalf of the PCCP Owner Societies)

pH Dependant Behaviour of Poly(Methacrylic Acid) Gels

The pH responsive gel used in the project is based on the monomer methacrylic acid (MAA). The gels are synthesised via a free-radical mechanism. The synthesis was carried out in an aqueous environment using 2,2'-azobis (2-methylpropionamidine) dihydrochloride (AMPA) and N,N'-methylene bisacrylamide (MBA); which are water-soluble free radical initiator and cross-link agent respectively (Scheme 1). The polymerizations were carried out in water (20%w/v MAA; 0.03%w/v MBA; 0.06%w/v AMPA) under a nitrogen atmosphere. The polymerization was initiated thermally at 70°C, and gelation occurred in less than one hour. Residual monomer was removed from the gels by alternate soaking in water and methanol.

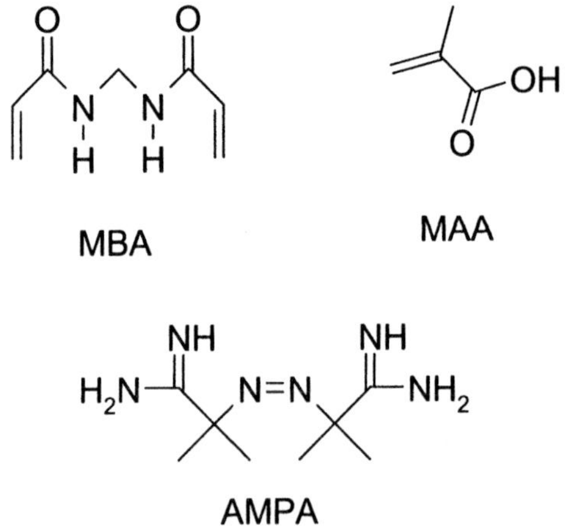

MBA   MAA

AMPA

Scheme 1. Reactants used for the synthesis of cross-linked PMAA gel.

The swelling of these gels were determined by soaking in solutions of 0.26M $KBrO_3$ adjusted to various pHs for 3 days, during which time the pH of the solutions values were monitored and any drift due to the gel corrected using dilute H_2SO_4 and NaOH solutions. The bromate was used to mimic the ionic strength of the oscillating reaction. The percentage increase in weight from the dry gel to the swollen gel was calculated at each pH, and the results are shown in Figure 2.

It is notable that at high pH the gel starts to collapse once more. This is due to charge screening by the Na^+ ion involved in adjusting the pH inducing further coiling. At low pH, the polymethacrylic acid (PMAA) chains form highly compact hydrophobic clusters, which are joined by short extended polymer chains. As the pH is increased the size of the clusters decreases up to a limit,

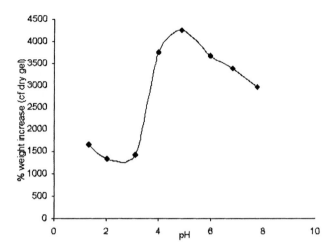

Figure 2. Equilibrium swelling of gel in a 0.26M KBrO$_3$ at various pH, adjusted using H$_2$SO$_4$ and NaOH solutions. .(Reproduced from reference 7 by permission of The Royal Society of Chemistry on behalf of the PCCP Owner Societies)

beyond which the clusters disintegrate completely to an expanded polymer chain. This phenomenon was observed and investigated by Turrob (*9*) and Bednář (*10*).

Force Production by PMAA Gel in pH Oscillator

In order to eventually be able to produce nanoscale machines and pumps, the swelling behaviour of the gel to be harnessed in order for mechanical work to be done. In order demonstrate the ability of the system to produce work, a modified version of the experimental set-up devised by Johnson-Kendall-Roberts (JKR experiment) (*11*) is used.

In order to give the gel a well-defined shape for the experiment, a mould was designed, which consisted of rubber with a circle cut out of it sandwiched between two glass slides and held in place by two spring clips (Figure 3). The cavity in the mould was 1mm thick and 5mm in diameter. Before use, the glass was treated with a solution of vinyltrichlorosilane for two hours, and then cleaned with pure methanol and pure water before being dried overnight in an oven at 80°C. This introduced pendant vinyl groups to the silicon surface so the gel, when polymerized in the mould, would be fixed to the glass surface.

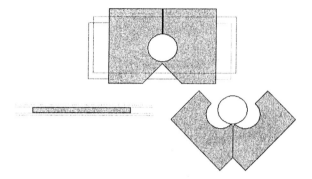

Figure 3. The mould design showing how the mould can be opened to leave the gel fixed to the glass slide.

The gel was made by pouring a degassed aqueous solution of monomer, initiator and cross-linker into the mould (MAA 47%w/v, AMPA 0.9%w/v and MBA 1.1%w/v). The mould was then placed in an oven at 80°C for 2 hours, before being placed in a water bath and the mould opened underwater. The gel, attached to the glass slides is stored under water.

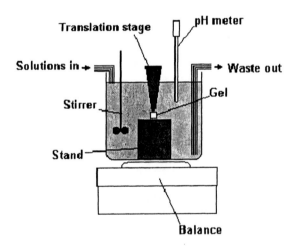

Figure 4. Schematic of the modified JKR experiment used to measure the force produced by the bulk PMAA gel as it changes volume in the oscillating reaction .

The force measurements were made in a modified JKR which consists of a 100ml glass beaker placed on an balance accurate to four decimal places. The sample is placed on a stage inside the beaker, and then a Z-translational stage, controlled by a worm microdrive, is lowered down onto the sample holding it in place (Figure 4). The solutions for the oscillating reaction are then pumped into the beaker with a peristaltic pump until the optimum level is reached, at which point waste is pumped out at an equal rate to maintain a constant volume. The reacting solution is then stirred using a paddle mounted on a 12-volt motor, at as slow a rate as possible whilst maintaining good mixing, in order to minimize noise in the readings. The concentrations of the reactants used were the same as used for the previous experiments.

As the gel swells and deswells in the oscillating solution, the translational stage is set in position when the pH is low and the gel contracted, so that it is applying a force in the order of a gram. Changes in this force were then monitored via the balance over the coarse of numerous oscillations (Figure 5). The pH data is presented as measured. The force data has been smoothed using a FFT smoothing function to remove only the stirring noise, as this has a constant and definite frequency.

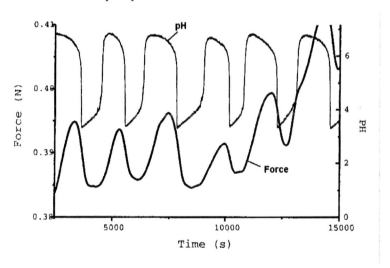

Figure 5. Chart displaying the variation in both pH and measured force over time as the hydrogel responds to the oscillating reaction . Residence time 1000s, at room temperature. Input concentrations: $[SO_3^{2-}] = 0.075M$, $[BrO_3^-] = 0.065M$, $[Fe(CN)_6^{4-}] = 0.02M$, $[H_2SO_4] = 0.01M$.

The oscillation appears to have a longer period than that seen in figure 1. It is not clear whether this is due to a buffering effect of the gel, or natural variation in room temperature. Although an attemp was made to keep the same residence time as in previous experiments, some mismatch may be present owing the use of different apparatus volumes, and this may also contribute to the change in oscillation period.

Over the four hour time period, the force produced can be seen to follow the oscillating reaction very closely, indicating that the diffusion through the gel is still quite fast relative to the timescale of the oscillations. The maximum force produced by the gel is 0.012N. The slight upward drift in the data is due to an increase in mass caused by a small mismatch between the flow rates of the pumps used.

Conclusions

A pH oscillator has been developed to cover the range 3 – 7 with a time constant of approximately 20 minutes. The reaction has been coupled with a responsive gel to generate a model system with spatio-temporal oscillations. In a previous paper these large dimensional changes have been observed by optical microscopy, and now have been harnessed mechanically to show evidence of the ability of these systems to generate force. Of course, the measured force is only a small potentially available force, as it is only measured in one dimension, but it displays that these systems can be used to directly convert chemical potential to mechanical work. Further work in this field will concentrate on more efficiently harnessing the systems, and to developing pH responsive valves so that devices such as pumps can be devised. In addition the fundamentals of this approach will be investigated by using Atomic Force Microscopy to look at the work generated by a single polymer chain changing it's conformation in response to the stimuli of changing pH.

The results of this initial study would indicate that the physical response of pH responsive polymers to changes brought about by non-linear kinetic systems, have great potential to be used to generate usable work in a new generation of 'soft' machines.

References

1. Dagani, D. *Chemical & Engineering News*, **1997**, June 9, pp 26-37.
2. Shiga, T. *Adv. Polm. Sci.*, **1997**, 131, p 134.
3. Beebe, D.J.; Moore, J.S.; Bauer, J.M.; Yu, Q.; Liu, R.H.; Devadoss, C.; Jo B.-H. *Nature*, **2000**, 404, pp 588-590.
4. Yoshida, R.; Tanaka, M.; Onodera, S.; Yamaguchi, T.; Kokufuta, E., *J. Am Chem* Soc., **2000**, 104, pp 7549-7555.

5. Yoshida, R.; Ichijo, H.; Hakuta, T. *Macromol. Rapid Comm.*, **1995**, 16, pp 305-110.

6. Rabai, G.; Epstein, I.R. *Acc. Chem. Res.*, **1990**, 23, pp 258-263.

7. Crook, C.J.; Smith, A.; Jones, R.A.L.; Ryan, A.J. *Phys. Chem. Chem. Phys.*, **2002**, 4, pp 1367-1369.

8. Edblom, E.C.; Luo, Y.; Orban, M.; Kustin, K.; Epstein, I.R.; *J. Phys. Chem.*, **1989**, 93, 2722-2727

9. Turro, N.J.; Caminati, G.; Kim, J. *Macromol*, **1991**, 24, pp 4054-4060.

10. Bednář, B.; Trnena, J.; Svoboda, P.; Vajda, S.; Fidler, V.; Prochazka, K. *Macromol*, **1991**, 24, pp 2054-2059.

11. Johnson, K.L.; Kendall, K.; Roberts, A.D. *Proc. R. Soc. London*, Ser. A **1971**, 324, p 301.

Chapter 7

Spontaneous Deformations in Polymer Gels Driven by Chemomechanical Instabilities

F. Gauffre, V. Labrot, J. Boissonade, and P. De Kepper

Centre de Recherches Paul Pascal (CNRS), Av. Schweitzer,
F–33600 Pessac, France

We show that a chemomechanical system can exhibit complex dynamical volume and shape changes in a steady homogeneous environment. This is experimentally achieved by coupling an acid autocatalyzed reaction (the chlorite-tetrathionate reaction) with cylinders of pH-responsive hydrogel. Although, in homogeneous conditions the reaction cannot produce oscillations, reaction-diffusion instabilities give rise to dynamical concentration patterns that induce large volume changes in the gel sample. In a last part, we elaborate on dynamical instabilities induced by the cross-coupling between geometrical changes in a piece of chemically responsive gel and a bistable reaction.

Responsive gels are networks of crosslinked polymers that can exhibit sharp changes in volume, mesh size, or transparency in response to an external stimulus (1). These materials can be made responsive to a wide variety of stimuli including temperature, electric or magnetic fields, or changes in solvent composition. They are considered as very promising for technological uses such as actuators (2), microvalves (3), or drug release devices (4). Recently, a few teams, inspired by the concepts of dissipative structures, have started to couple chemically responsive-gels to chemical reactions maintained in non equilibrium conditions. Thus, sustained oscillations of a gel could be obtained by coupling a pH-responsive gel to a pH-oscillating reaction (5). Borkmans et al. have given a theoretical description of the dynamics of an oscillating model reaction within a swelling-deswelling gel, and shown that the oscillations can be shifted to a control parameter domain where the reaction is stationary otherwise (6). Siegel et al. have designed a very elegant device in which the coupling between a non-oscillating enzymatic reaction and the permeability of a pH-responsive membrane of gel induces an oscillatory instability (7). Today, it is still an important challenge to find chemomechanical systems that can produce rhythms and forms by coupling chemical processes with mechanical processes. In this paper, we propose a new experimental approach by coupling an acid autocatalytic reaction to a pH-responsive gel.

The reaction-diffusion dynamics of the acid autocatalytic Chlorite-Tetrathionate (CT) reaction was thoroughly investigated (8). Like other autocatalytic reactions, the CT reaction exhibits a more or less long induction period followed by a rapid switch to thermodynamic equilibrium. In a continuous stirred tank reactor (CSTR), this reaction can exhibit bistability. One state is obtained at high flow rates or at highly alkaline feed flows, when the induction time of the reaction is much longer than the residence time of the reactor. The reaction mixture then remains at a very low extent of reaction and this state is often named the "Flow" (F) or the "Unreacted" state. In our experimental conditions, the F state is alkaline (pH $\approx$10). The other state is obtained for low flow rates or for weakly alkaline feed flows, when the induction time of the chemical mixture is shorter than the residence time of the reactor. It is often called a "Thermodynamic" (T) or "Reacted" state because the reaction is almost completed in the CSTR. In our experimental conditions, the T state is acidic (pH $\approx$2). The domains of stability of these two states overlap over a finite range of parameter.

When a piece of an inert gel or of an other porous material is immersed in a CSTR maintained in the F state, two types of concentration profiles can develop in the depth of the gel. If the diffusive exchanges between the gel and the reactor contents occur over a time shorter than the induction time of the reaction, the reaction mixture remains almost unreacted everywhere in the gel, so that the concentration profiles are almost flat and the whole gel is alkaline. On the contrary, if the diffusive exchanges occur over a time larger than the induction time, then the reaction "has time" to occur within the gel. Because the reactive medium at the boundary of the gel is maintained in the F state, a stable reaction front usually forms separating two areas of dramatically different chemical compositions. Close to the interface the extent of the reaction is low and the solution is alkaline, whereas the extent of the reaction is high and the pH is acidic beyond the front. This new spatial state in the gel is refered to as the "mixed" (FT) state.

There exist a finite range of gel sizes over which both profiles are simultaneously stable. This phenomenon is named "spatial bistability" (9). In addition, a new type of reaction-diffusion instability was evidenced when studying the CT reaction. Indeed, although it cannot oscillate in homogeneous conditions, the CT reaction exhibits complex dynamical behaviors, including trigger waves and oscillations (10–12). Unlike spatial bistability which directly stems from the autocatalytic kinetics of the reaction, the oscillations and excitability originate not directly in the local kinetics mechanism but in a difference in the values of the diffusion coefficients of the species. In this case H^+, which is the activatory species of the reaction, has an effective diffusion coefficient about three times larger than the other species. As a consequence, the diffusive exchanges between the gel and the reactor contents occur over different time scales for the activator and for the other species which gives rise to this temporal instability (11).

The CT reaction was chosen as a model system for coupling with pH-responsive hydrogel because: (i) pH difference between the reacted and the unreacted state is very large (ΔpH ≈ 8) so that it can readily induce volume changes in conventional pH-responsive gels; (ii) its chemical state depends on the size of the system, so that volume changes might introduce a feedback on the chemistry; (iii) it leads to oscillatory and excitability phenomena in a gel under fixed boundary conditions.

Spontaneous Deformations of pH-responsive Gels Coupled to the CT Reaction

Materials and Methods

Synthesis of hydrogels. Hydrogels were prepared by free radical copolymerisation of acrylic acid (AAc) and N-isopropylacrylamide (NI-PAAm) in dioxane, using 2,2-azobis(isobutyronitrile) (0.2 mol %) as the initiator and N,N'-methylenebisacrylamide (0.8 mol %) as a cross-linker. The molar composition of AAc in the monomer mixture was 5 % and the total monomer ratio of the cross-linked hydrogel was 50 wt %. Before synthesis, N-isopropylacrylamide (ACROS organics) and 2,2-azobisisobutyronitrile (FLUKA) were recrystallized respectively in toluene and methanol, and AAc (AVOCADO) was purified by cryodistillation over CaH_2. N,N-methylenbisacrylamide (FLUKA) was used as received and dioxane (SDS) was dried on activated molecular sieves. Gelation was carried out in cylindrical silicone tubes with internal diametersd varying from 0.5 to 1.7 mm (and length around 30 mm) under a nitrogen atmosphere at 75°C for 24h. The hydrogels were removed from their molds, washed in 100/0 and 50/50 vol % methanol/water mixtures for 1 day each, and eventually in an aqueous chloroform solutions (1 vol %) during at least 3 days before use. The copolymer composition of the network was determined by conductimetric titration of a suspension of pulverized dried gel. It was found that the AAc molar percentage is equal to $10 \pm 1\%$. This is twice the ratio of the initial monomer mixture.

Swelling measurement. Cylindrical gel samples were synthesized in 1mm diameter molds. After the washing procedure, the reference length ($L_0 \approx 30$mm) of the cylinders in pure water and at T=25°C was measured. The pH dependent swelling measurement was performed in a flow cell system using the same reactor as for the coupling experiments (see below). The reactor is kept at 35°C and the pH is cycled (up and down) by flowing with different mixtures of two 0.5M NaCl aqueous solutions: neutral and alkaline (NaOH) or neutral and acidic (HCl). For each pH values we waited for equilibration during approximately 3 hours before measuring the gel length L.

Chemomechanical coupling. We have used a home made CSTR (volume 46 ml) which allows to maintain a cylindrical piece of deformable gel in an open bath of reactants (Figure 1). To insure a good mixing along the revolution surface of the cylindrical gel and a relative stabilization in position of this gel, the CSTR was made of three interconnected vertical chambers of cylindrical shape. A turbine-like magnetic stirrer located at the bottom of the central chamber creates a hydrodynamic vortex and pumps the fluid down; then the fluid is pushed back to the top of the middle chamber through the side connections. This introduces a strong mixing of the bulk contents with a recirculation time of approximately 2 seconds. Sodium chlorite (Prolabo,96% pure) and potassium tetrathionate (Fluka) were used without further purification. The feeding solutions were stored in four separated reservoirs containing respectively an alkaline sodium chlorite solution ($[NaClO_2] = 2 \times 10^{-1}$ M, [NaOH] =1.5×10^{-4} M), an alkaline potassium tetrathionate solution ($[K_2S_4O_6] = 5 \times 10^{-2}$ M, [NaOH] =1.5×10^{-4} M) and two sodium hydroxide solutions of different concentrations. The solutions were pumped by precision piston pumps (pharmacia P500) and premixed just before being injected into the CSTR at the level of the turbine. The flows of chlorite and tetrathionate solutions were respectively maintained at 47 ml/h and 53 ml/h, and the sum of the two sodium hydroxide solutions at 50 ml/h. The ratio between the flows of these two sodium hydroxide solutions were varied to tune the total amount of sodium hydroxide introduced in the CSTR. The total feed flow was 150 ml/h which corresponds to a residence time of 18.5 minutes for the CSTR. The cylinder of gel was glued at one end on a Plexiglas stopper and introduced at the top of the

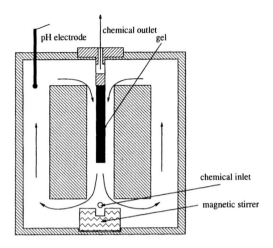

Figure 1. Schematic cross-section of the CSTR with the gel cylinder (black).

central chamber (of inner diameter 15 mm). The CSTR was set into a thermostated bath at 35°C so that the gel is clear in the swollen state and turbid in the collapsed state (see next section). The pH into the CSTR was monitored using a pH electrode, and the gel state was recorded on a time lapse VCR using a black and white camera. We have filmed either directly the gel across the plexiglass wall of the reactor or using a shadow-graph technique. Shadowgraph projections allow to produce "shadows" of the contours of the gel and to view any sharp gradient of refraction index inside the gel. This enables to clearly follow the local diameter changes.

Swelling Behavior of pH-Responsive Gels

Pure poly(NIPAAm) hydrogels are known to be thermoresponsive and to undergo an almost continous first-order volume transition from swollen to shrunken state by increasing temperature (*13*) above the lower critical solution temperature (LCTS≈33°C). The introduction of acidic co-monomers, for instance AAc monomers, makes the gel pH-responsive (*14*). Indeed, in alkaline conditions these monomers become charged, which induces a swelling driving force on the network due both to the repulsive electrostatic interaction between charged groups and to the osmotic pressure from the free counter ions (*14*). The numerous experimental works reported in the litterature, show that some features of the swelling-deswelling processes, such as transition thresholds, amplitude and kinetics, depend on the chemical composition, microscopic structure and shape (*15*) of the gels, so that one cannot foresee the macroscopic properties of a gel from the synthetic procedure. Only an *a posteri* characterization of the material is reliable.

We have first studied the temperature dependence of the linear swelling ratio L/L_o of poly(NIPAAm-co-AAc) gel cylinders at different pH values.

It was observed that a maximal volume variation between $pH = 2$ and pH $= 10$ can be obtained for temperatures close to $T = 33°C$. Furthermore, above this temperature, collapsed gels are turbid which provides an easy way to distinguish between the shrunken and swollen parts within the piece of gel. As a consequence, we have set the experiment temperature at $35°C$ for chemomechanical studies. Figure 2 shows the equilibrium pH dependence of the linear swelling ratio L/L_o of gel cylinders at $T = 35°C$. Experiments are operated in 0.5M NaCl solutions in order to be comparable to the high ionic conditions of the chemomechanical coupling experiments. The swelling ratio increases with pH with a marked transition around $pH = 5.3$. No hysteresis can be observed within our experimental accuracy. For values below $pH = 5$ the samples are turbid, this is thought to result from microphase separation (16).

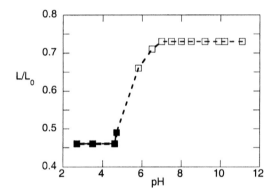

Figure 2.   pH dependence of the linear swelling ratio L/L_0, for poly(NIPAAm-co-AAc) gels at 35°C in 0.5M NaCl. Black symbols stands for turbid samples.

Chemomechanical Coupling

The coupling is experimentally achieved by immersing a cylinder of pH-responsive gel in a CSTR where the CT reaction is operated (Figure 1). The CSTR contents is kept in the F (alkaline) state so that it differs only slightly from the input composition. We use the value of the sodium hydroxyde concentration $[OH^-]_0$ in the total input flow as control parameter. When a piece of gel previously soaked in a neutral or an alkaline solution is introduced in the CSTR, it remains clear and swollen, whatever the value of the control parameter. To start the experiment, a cylinder of gel is soaked into an acidic solution ($pH = 2$) outside the CSTR, which makes it to shrink. During the same time, the CSTR is fed with reactants until the desired asymptotic state is reached. Then, the piece of gel is immersed into the CSTR and after a transient, the gel reaches an asymptotic state. The feed concentration $[OH^-]_0$ is then progressively varied to investigate other values of the control parameter, and for each value we waited at least five hours for the asymptotic state to settle down. In typical experiments we observe the following sequence:

For the highest $[OH^-]_0$ values, the gel swells to a completely transparent swollen state. In this state no further dynamical change is observed.

Setting the control parameter to lower $[OH^-]_0$ values leads to a new asymptotic regime characterized by periodic turbid waves propagating in a completely transparent gel (Figure 3). The waves spontaneously emerge from the top of the cylinder, where it is glued to the holder, and travel downwards with a speed of the order of 1cm/hr. Increasing the value of $[OH^-]_0$ leads to slower waves with a shorter tail. The wave propagation is accompanied by a bottleneck deformation of the cylinder which, in some cases, is very narrow. A bend can also form and move along with the location of the turbid wave.

For still slightly lower $[OH^-]_0$ values, the system reaches a complex dynamic regime in which the gel is essentially collapsed with a turbid core surrounded by a thin transparent shell. Some sections of the gel start to swell and shrink back at random positions as shown on Figure 4 for an area located at mid-height of the cylinder. Like in the previous case, bends often form at the location of strong gradients of thickness.

For the lowest $[OH^-]_0$ values, the gel reaches a collapsed stationary state with a turbid core and a thin transparent shell. Snapshots taken during the transient propagation of the collapsed state into the swollen state are shown in Figure 5. The length of the gel cylinder in the collapsed state is typically 40% of the maximal length in the swollen state. When the gel is collapsed, a "bubble" structure is observed as in Figure 5 (bottom).

The previous procedure was repeated with gels made with different mold diameters d in order to build a phase diagram which is a representation of the non equilibrium asymptotic states in the ($[OH^-]_0$-d) plane (Figure 6). The diagram is limited on the left-hand-side by the stability limit of the F state in the CSTR at $[OH^-]_0 = 4 \times 10^{-3}$ M. Beyond this limit, any perturbation induces a fast transition of the CSTR contents into the T (acidic) state. On the right-hand-side, the sequence of behaviors that was previously described is observed at all sizes. The range over which this sequence spreads decreases with the initial diameter of the cylinder.

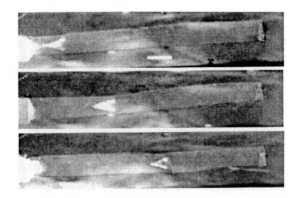

Figure 3. Turbid trigger wave in a pH responsive gel cylinder. The photographs are tilted by 90°. Snapshots from top to bottom at 0, 110, 220 minutes. Experimental conditions: $d=1mm$, $[OH^-]_0=5.3\times10^{-3}M$. White scale bar=3mm.

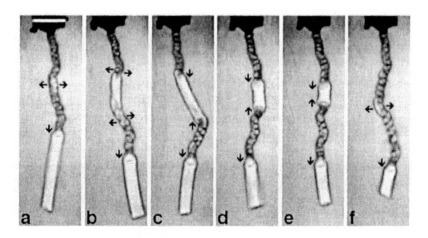

Figure 4. Shadowgraph snapshots of the dynamical shape changes in the oscillatory regime. Arrows point in the direction of the main swelling-deswelling dynamics. Snapshots from (a) to (f) at 0, 20, 30, 40, 44, 60 minutes, cover about one period of oscillation of the mid-heigth point of the gel. Experimental conditions: $d=0.5mm$, $[OH^-]_0=4.25\times10^{-3}M$. White scale bar in (a)=3mm.

Discussion

Most of the observed behavior can be explained by simply assuming that the volume (size) changes of the gel are only slaved to the acid-base patterns, that result from the coupling of the autocatalytic reaction with a differential diffusion process of reactive species in the gel (*11*). One can qualitatively map the presently reported sequence of regimes in the ($[OH^-]_0$-d) plane (stationary collapsed → oscillatory → trigger waves → stationary swollen) with the sequence of states (stationary FT → oscillatory FT → excitable F → stationary F) observed in non responsive gels (*11,12*). Also, as in the non responsive gels, we observe bistability between the swollen (resp. F) and the collapsed (resp. FT) states. In both cases, the stability range of the unreacted state (swollen or F) fully overlaps the stability range of the reacted state (collapsed or FT), so that the system cannot spontaneously switch from the unreacted to the reacted state. In this "spatial bistability" range the actual asymptotic state depends on whether or not a sufficient amount of acid is initially present within the gel. At fixed feed, there is a minimum diameter d_{min} below which the FT state of the gel is unstable and switches to the F state (*11,12*).

Note that, in the present case, trigger waves and oscillations exist only in a narrow range of parameter. The reason for this, is that the polymer network bears acidic functions in order to make it *p*H-responsive. Introducing immobile proton binding sites into the gel slows down the effective diffusion of protons and, as a consequence, reduces the range of parameter over which the excitability and oscillations are observed (Szalai, I. Private communication).

The present observation of periodic waves in the swollen gel require some further explanations. In fact, gluing the cylinder to the holder introduces a local impediment to the feed of the gel. This acts as a local effective increase in gel size where an acid bubble can be trapped and act as a pacemaker for the waves. Experiments on conical gels show that such an increase in diameter can lead to a periodic wave source (*12*). The propagating short turbid wave in Figure 3 is associated to an acidic pulse of approximately the same size. The acid front (not visible here) very slightly precedes the head of the turbid wave. The diameter drop at the location of the wave can be as large as a factor 2.5, that is a volume collapse is of one order of magnitude. Deformation waves often go with sharp bends and twists of the gel cylinder, which is thought to be due to local inhomogeneities.

Finally, we would like to point out that although the static and dynamic behavior of the responsive gel seems to follow the dynamical instabilities driven by the reaction-diffusion processes, the large size changes of the gel can also apply a feedback over the chemical state of the system. Indeed, previous experiments evidenced that in our range of diameter (0.5-3mm)

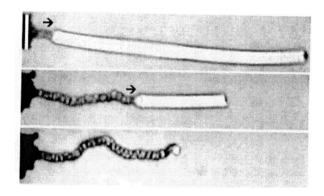

Figure 5. Shadowgraph snapshots of an initially swollen gel collapsing after an acidic perturbation. From top to bottom, time is: 0; 126; 350 minutes. Experimental conditions: d=0.5mm, [OH⁻]₀=4.25×10⁻³M. The photographs are tilted by 90°. Vertical scale bar=3mm.

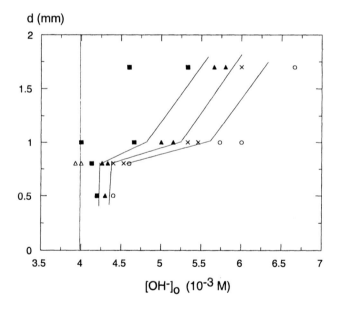

Figure 6. Non equilibrium phase diagram of the pH responsive gels coupled to the CT reaction. ○ swollen state; × trigger waves; ▲ volume oscillations; ■ collapsed state; △ transition of the CSTR into the T state. The parameter d is the inner diameter of the mold.

the chemical state of the gel is size dependent (11,12). Let us now describe how this size change could feed into our observed dynamics. Suppose we start from a swollen gel cylinder, the diameter of which is a little bit larger than d_{min}, the stability limit for the FT state in the gel. If we now apply an acidic perturbation to the piece of gel the system switches to the stable mixed FT state and as we have seen, the diameter of the gel dramatically shrinks. If the diameter becomes smaller than d_{min}, the FT state becomes unstable and the contents of the gel switches back to the akaline F state, where the gel swell and remains swollen if no other perturbation is done. Thus, because of the size change the system can behave as an excitable system even in the absence of any diffusion-driven instability. At this stage of our experiments, it is difficult to separe the contribution of these two possible sources of dynamical instabilities.

Oscillations Induced by Reaction and Size Changes Cross-Coupling

In a recent numerical study (17), spontaneous chemomechanical pulsations were achieved by coupling a mechanically responsive sphere of gel with a chemical reaction which cannot produce oscillations by itself. The toy model used in this study is based on a two variables model of a quadratic autocatalytic reaction described by the following reaction-diffusion system:

$$\partial u/\partial t = -u^2 v^2 + \nabla^2 u \qquad ; \qquad \partial v/\partial t = (12/7)u^2 v^2 + \nabla^2 v$$

To avoid oscillating instabilities, the diffusion coefficients of all species where set equal (11). The variables u and v are fixed to constant values $u_0 = 1$ and $v_0 = 0.05$ at the interface between the gel and the surrounding medium. When this system is solved numerically in a sphere of a given radius R with a given set of initial conditions, it leads to a stationary concentration profile along the radius. Two types of profiles can be obtained which correspond either to the "reacted" or "unreacted" states. Also, a domain of bistability between those two profiles is observed over a finite range of R. No oscillations nor excitability can result from this model.

Now, let us assume that the local swelling state of the gel depends on v. This dependence is introduced in the free energy of the gel through the classical Flory interaction parameter $\chi = \chi(v)$, which characterizes the energy term in the mixing chemical potential $\Delta\mu_{mix} = RT(\phi + ln(1 - \phi) + \chi\phi^2)$ where ϕ is the volume fraction of the polymer. The function $\chi = \chi(v)$ is taken as an increasing sigmoidal function of v that saturates at large v, when all the sensitive network sites are activated or deactivated. The generalized chemical potential describing the equilibrium properties of the network is the sum of the mixing and elastic terms: $\Delta\mu = \Delta\mu_{mix} + \Delta\mu_{el}$. The elastic term is taken as $\Delta\mu_{el} = K_{net}RT[(\phi/\phi_0)^{1/3} - \phi/(2\phi_0)]$ where the constant K_{net} depends on the properties of the network and ϕ_0 is the

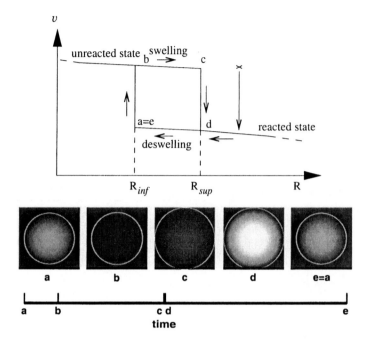

Figure 7. Schematic representation of the oscillation mechanism and snapshots of a cross-section of the sphere of gel during one period of oscillations. The grey scale maps concentration v (maxima: white; minima: black).

volume fraction in a reference state. The swelling process can then be numerically calculated, using the Maxwell-Stefan approach which asserts that, for moderate swelling ratio, each point of the network is submitted to a driving force linearly depending on the gradient of chemical potential : $F_{tot} = -\nabla(\Delta\mu)$, is equal to a friction force applied to the solvent molecules by the network: $F_{fr} = RT\phi G(\phi)(\xi_s - \xi_p)$, where ξ_s and ξ_p are the velocities of the solvent and the polymer respectively and $G(\phi)$ is a function describing the dependence of friction with polymer density. Details on these functions are reported with quantitative computations in a forthcoming publication (*17*). Here, we only report a series of visual snapshots of the state of the sphere corresponding to a typical case. They come with an explicative scheme of the oscillation mechanism described below.

When the above reaction-diffusion system is operated in such a responsive medium, cross-coupling between the volume changes and the reaction occurs, that can give rise to relaxation oscillations in both the radius of the sphere and the chemical composition. The oscillating process can be schemed as follows (Figure 7). If we start with a swollen gel with $R>R_{sup}$, then the unreacted state is unstable, and the gel starts to shrink. When R reaches R_{inf} (point a) , the reacted state becomes unstable, the system quickly switches to the unreacted state (b) and the gel starts to swell until it reaches $R=R_{sup}$ (c) and goes back on the reacted state (d). This process then repeats indefinitely. Note that if we call R_{max} the radius of the gel in a completely swollen state (corresponding to a uniform $v=0$ in the gel) and R_{min} the radius of the gel in a completely shrunken state ($v=1$ in the gel), then it is necessary to have $R_{max} >R_{sup}$ and $R_{min} <R_{inf}$ for the oscillations to occur. In other words, it is necessary that volume changes of the gel drive R beyond the limits of the bistability domain. Snapshots of a cross-section of the sphere during one period of the oscillations as given by the numerical model are displayed in Figure 7. The white circle represents the gel boundary, and the grey-scale maps to v (white for high v, black for low v). Snapshot (a) is taken for $R=R_{inf}$, just before the switch to the un-reacted state. Snapshot (b) is taken just after the switch. (c) corresponds to the maximum of swelling ($R=R_{sup}$) just before the switch to the reacted state, and (d) is taken shortly after the switch.

Note that the above mechanism cannot be fulfilled in our experiments, because the induction time of the reaction is quasi-infinite so that the upper limit of the bistable domain, R_{sup}, takes a quasi-infinite value (10–12). Then it would take a piece of gel of an infinite size to have the reaction switch spontaneously from the unreacted to the reacted state, and the os-cillation loop cannot be closed. Work is now in progress in order to devise an experimental system producing oscillations through this mechanism.

Conclusion

We achieved the coupling between a non-oscillating autocatalytic reac-tion and a pH-responsive gel in an open system. The cylindrical pieces of gel, immersed in an homogeneous and stationary reactive medium, ex-hibit large and complex dynamical volume and shape changes although the reaction itself cannot oscillate in homogeneous medium. Large amplitude deformations (a few milimeters in the diameter and a few centimeters in the length) are observed. It is believed that the mechanical response of the gel is slaved to pH spatial and temporal patterns resulting from pure reaction-diffusion instability. However, the chemical state of the system depends on the radius of the cylinder, so that changes in size are bound to feedback onto the reaction processes. Some of the consequences of cross-coupling mechanical changes and reaction-diffusion processes are discussed and exemplified in a short report on a recent numerical work showing that periodic volume pulsations can be obtained by applying a size-dependant feedback over the hysteresis loop of a bistable chemical system.

We thank I. Szalai, E. Dulos, P. Borckmans, G. Dewel, A. De Wit and K. Benyaich for providing useful informations and also stimulating discussions.

References

1. DeRossi, D., Kajiwara, K., Osada, Y., Yamauchi, A. Eds. *Polymer gels;* Plenum: New York, 1991. Osada, Y.; Ross-Murphy, S. *Sci. Am.* **1993**, *268*, 82–87.

2. Shahinpoor, M.; Bar-Cohen, Y.; Simpson, J. O.; Smith, J. *Smart Mater. Struct.* **1998**, *7*, R15–R30.

3. Beebe, D. J.; Moore, Jeffrey, S.; Bauer, J. M.; Yu, Q.; Liu, R. H.; Devadoss, C.; Jo, B.-H. *Nature* **2000**, *404*, 588–590.

4. Bae, Y. H.; Kwon, I. C.; chapter 4.*Stimuli-Sensitive Polymers for Modulated Drug Release;* Academic Press, 1998; pp 93–134.

5. Yoshida, R.; Takahashi, T.; Yamaguchi, T.; Ichijo, H. *J. Am. Chem. Soc.* **1996**, *118*, 5134–5135. Yoshida, R.; Tanaka, M.; Onodera, S.; Yamaguchi, T.; Kokufuta, E. *J. Phys. Chem. A* **2000**, *104*, 7549–7555. Crook, C. J.; Smith, A.; Jones, R. A. L.; Ryan, A. J. *Phys. Chem. Chem. Phys.* **2002**, *4*, 1367–1369.

6. Borkmans, P.; Benyaich, K.; De Wit, A.; Dewel, G. This volume.

7. Dhanarahajan, A. P.; Misra, G. P.; Siegel, R. A. *J. Phys. Chem. A.* **2002**, *106*, 8835–8838.

8. Tóth, A.; Lagzi, I.; Horváth, D. *J. Phys. Chem.* **1996**, *100*, 14837. Tóth, A.; Horváth, D.; Siska, A. *J. Chem. Soc. Faraday Trans.* **1997**, *93*, 73–76. Horváth, D.; Tóth, A. *J. Chem. Phys.* **1998**, *108*, 1447–1451. Tóth, A.; Veisz, B.; Horváth, D. *J. Phys. Chem. A* **1998**, *102*, 5157. Fuentes, M.; Kuperman, M. N.; De Kepper, P. *J. Phys. Chem. A* **2001**, *105*, 6769–6774.

9. Blanchedeau, P.; Boissonade, J.; De Kepper, P. *Physica D* **2000**, *147*, 283–299.

10. Boissonade, J.; Dulos, E.; Gauffre, F.; Kuperman, M. N.; De Kepper, P. *Faraday Discussions* **2001**, *120*, 353–361.

11. Fuentes, M.; Kuperman, M.; Boissonade, J.; Dulos, E.; Gauffre, F.; De Kepper, P. *Phys. Rev. E* **2002**, *66*(056205).

12. Gauffre, F.; Labrot, V.; Boissonade, J.; De Kepper, P.; Dulos, E. *J. Phys. Chem. A,* **2003**, *107*, 4452–4456.

13. Hirokawa, Y.; Tanaka, T. *J. Chem. Phys.* **1984**, *81*, 6379–6380.

14. Shibayama, M.; Ikkai, F.; Inamoto, S.; Nomura, S.; Han, C. C. *J. Chem. Phys.* **1996**, *105*, 4358–4366.

15. Hirotsu, S. *Macromolecules* **1992**, *25*, 4445–4447.

16. Borue, V.; Erukhimovich, I. *Macromolecules* **1988**, *21*, 3240–3249.

17. Boissonade, J. *Phys. Rev. Lett.* **2003**, *90*,188302.

Chapter 8

Pattern Formation in a Complex Reaction System by Chemomechanical Coupling

Stefan C. Müller

Institut für Experimentelle Physik, Otto-von-Guericke-Universität
Magdeburg, Postfach 4120, D–39016 Magdeburg, Germany

In an experimental study of the polyacrylamide-methylene
blue-sulfide-oxygen (PA-MBO) sytem a variety of concentration patterns have been observed. Initially, stripes or
hexagons form. The latter transform to chevrons or white-eye
patterns. Honeycombs are a third type of structure that develops from either white-eye or hexagonal patterns. For
hexagons, the pattern wavelength shows a nearly proportional
dependence on the height of the reaction layer. The motion of
small tracer particles serve to monitor physical changes within
the medium. There are three distinct phases of particle movement. The first one is caused by hydrodynamic flows in the
pre-gel solution, whereas the third, and possibly the second,
phase arises from localized volume changes within the gel.

Introduction

During the last decades the investigation and analysis of concentration patterns forming in chemical non-equilibrium systems has attracted considerable attention. An overview of many interesting aspects of this field of research can be found in Ref. 1. The macroscopic order of these systems often arises from the diffusive coupling of nonlinear reaction processes which can lead to the formation of stationary Turing patterns (*2,3*) or propagating waves (*4,5*). Turing patterns form in certain reaction-diffusion systems in which the inhibitor species diffuses faster than the activator species. An important motivation for their investigation is the possible involvement in biological morphogenesis (*6*). There is some evidence that concentration patterns of this kind can also generate shape and geometry by mechanical deformations. In 1996, the first experimental example in which propagating chemical waves lead to a periodic swelling of a gel matrix was reported (*7*). Another system in which these effects may be more profound is the polyacrylamide methylene blue-oxygen-sulfide (PA-MBO) system (*8*).

The oscillating MBO (*9,10*) system involves the methylene blue catalyzed oxidation of HS^- by molecular oxygen in an aqueous medium at $pH \approx 12$. The monomer methylene blue exists in the blue oxidized form MB^+, the colorless reduced form MBH and the $MB^\bullet$ radical, an intermediate oxidation step. Under

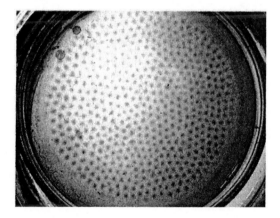

Figure 1. Hexagonal pattern in the PA-MBO system. Thickness of reaction layer, 1.7 mm; diameter of petri dish, 7 cm.

atmospheric conditions only the MB$^+$ ion is stable, whereas MBH and the radical are oxidized by the oxyen of the air. In the PA-MBO system, the MBO reaction is started in the presence of a polymerizing acrylamide/bisacrylamide pre-gel solution. In the course of several minutes, this complex reaction system forms transient hexagons, stripes, and zig-zag patterns that have typical wavelengths of 1-3 mm as reported by Watzl and Münster (8). An example of a hexagonal pattern is presented in Figure 1. Evidence for more complex patterns in this reaction system has been provided in Refs. 11-13 in an investigation that focused on later stages of pattern evolution. Some examples of this study are shown in Figure 2.

The geometry and orientation of the fundamental stripe and hexagon patterns can be influenced by electric fields and by light (14), which is in qualitative agreement with results from reaction-diffusion models (15), and should support the hypothesis that the patterns primarily occur from Turing instabilities. However, in 1998, Kurin-Csörgei et al. (16) presented experiments in which neither methylene blue nor sulfite was required for patterns to develop and they conclude that patterns form due to a hydrodynamical instability during the polymerization and gelation process. Subsequently, a model was suggested

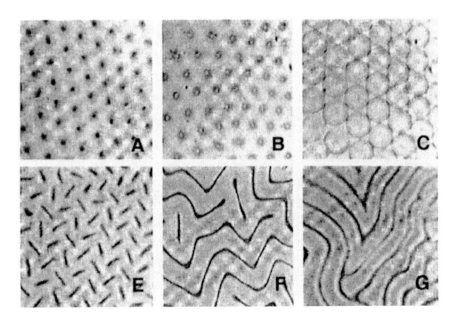

Figure 2. Patterns in the PA-MBO reaction obtained for different initial concentrations of sodium sulfide: (A) 32.7 mM, (B) 38.1 mM (early pattern), (C) 38.1 mM (late pattern), (D) 40.0 mM, (E) 41.4 mM, and (F) 42.1 mM. Field of view, 16 × 16 mm^2. (Reproduced with permission from reference 12. Copyright 2000 Wiley-VCH Verlag.)

that is capable of generating Turing patterns in the absence of methylene blue and explains the patterned swelling of the gel matrix, due to an inhomogeneous spatial distribution of molecular oxygen (*15*).

There is now clearly the need for additional investigations of this system in order to unravel the underlying mechanisms that may lead to a satisfactory explanation of the complexity of the observed patterns. In the following we present experimental data on the evolution of hexagonal patterns that contribute to the possible role of chemo-mechanical coupling in this system.

Experimental

Following Ref. 8, we use the complete PA-MBO system, adding sodium hydroxide to further increase its pH value. The gel is left uncovered and kept under air at room tempreture. Consequently, the reaction system is semi-closed in the sense that the reactant oxygen is supplied from the air, whereas sulfide is continuously consumed. A petri dish (diameter: 7.0 cm) serves as the reaction container for the PA-MBO medium. The blue patterns on a colorless background can be seen with the naked eye and are recorded by two-dimensional spectrophotometry for quantitative analyses.

Preparation of the system: 0.24 ml N,N'-methylene-bisacrylamide (2 g / 100 ml H_2O), 0.13 ml NaOH (0.5 M), and 0.25 ml triethanolamine (30 g / 100ml H_2O) are added to 3.17 ml acrylamide (20 g / 100ml H_2O). A mixture of 0.29 ml methylene blue (0.318 g / 100 ml H_2O), 0.5 - 0.65 ml sodium sulfide (3.902 g Na_2S / 100ml H_2O), and 0.87 ml sulfite solution (Na_2SO_3; 0.252 g / 100 ml H_2O) is added. Before starting the polymerization with 0.13 ml ammonium persulfate (5 g / 25 ml H_2O), double-distilled water is added to reach a final volume of 7.4 ml. Prior to all experiments, sodium sulfide is purified by refluxing commercially available Na_2S with Cu powder in ethanol under a nitrogen atmosphere for 1 h.

Results and Discussion

Height-Dependence of Pattern Wavelength

For reaction conditions with low initial $[Na_2S]_0$ (30-40 mM), the PA-MBO system self-organizes patterns of blue dots (i.e., high concentration of MB^+) that are arranged on hexagonal lattices (cf. Figure 2A). Together with irregular stripes and zig-zags (Figures 2E,F), hexagons are the fundamental patterns that develop during the early stages of the reaction. Moreover, hexagons exist in well-gelled media, whereas stripes and zig-zags are typically found in highly viscous, but not fully gelled systems ($[Na_2S]_0 > 40$ mM). Our optical detection revealed no convincing evidence for hexagonal structures prior to gelation.

The symmetry of the hexagonal patterns is typically affected by a small number of mismatches, such as penta-hepta defects. For a quantitative analysis, we calculated the two-dimensional power spectra from the raw data of individual video images. In this analysis, the digitized image data is Fourier-transformed from the (x,y)-space into the wave vector space (k_x, k_y). The inset of Figure 3A shows an example for the power distribution of a typical hexagonal PA-MBO pattern revealing six dots of high signal power that are arranged at a regular spacing on a circle. The radial distribution of the signal power in the main plot of Figure 3A shows a dominant wave vector at $k = 0.5$ mm^{-1}. The corresponding wavelength of 2.0 mm denotes the average, minimal spacing of MB$^+$ dots in the real-space pattern.

Hexagonal structures were investigated for reaction systems with layer heights in the range of 1.35 mm to 4.0 mm. Systems with thinner reaction layers always failed to generate optically detectable structures, which suggests the existence of a critical layer height of approximately 1 mm below which no patterns are formed. Visual inspection of the pattern shows that the wavelength of the structure increases with increasing layer heights, whereas their overall symmetry and appearance remains unchanged.

The change in pattern wavelength with layer height was quantified by a power spectra analysis. The corresponding data in Figure 3B were obtained from experiments, in which the reaction systems were subjected to two different light intensities. Apparently, the pattern wavelength is nearly proportional to the layer height regardless of the intensity of the applied illumination. Up to now, we have not attempted to investigate patterns for layer heights above 4 mm. Because they will require the use of overall larger reaction vessels that assure a sufficiently large aspect ratio between the vertical and horizontal dimensions of the system.

Analysis of the Motion of Colloidal Particles

The inspection of the reaction system with an optical microscope reveals the presence of small particles that carry out a systematic motion within the medium. These particles (diameter ≈ 30 μm) cannot be readily dissolved or broken apart by treatment in an ultrasonic bath. It is likely that they are similar or identical to colloidal sulfur that forms in the reaction. The impact of these colloidal particles on the reaction dynamics in

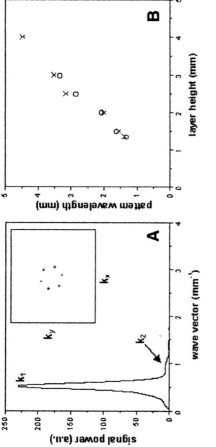

Figure 3. *(A) Power spectrum of the signal intensity as obtained from a typical hexagonal pattern in the PA-MBO medium. The dominant wavelength is $\lambda_1 = 1/k_1 = 2.0 \pm 0.3$ mm. The inset shows the square modulus of the pattern's two-dimensional Fourier transform. (B) Wavelength of hexagonal patterns vs. height of the reaction layer under white-light illumination of 17 W/m^2 (circles) and 30 W/m^2 (crosses). (Adapted with permission from reference 13. Copyright 2002 Oldenbourg Wissenschaftsverlag.)*

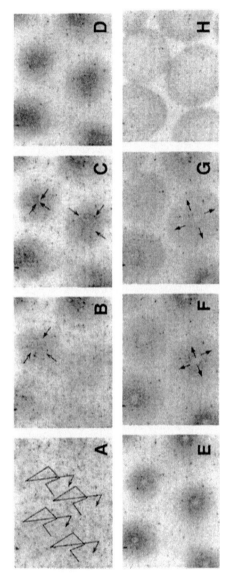

Figure 4. *Sequence of snapshots: frame A shows a homogenous phase (note that the black speckles are talcum particles used to trace motion on the gel). Transition to hexagonal patterns is shown in frames B-D, followed by the formation of 'white-eye' structures in E. Frames F-H show the subsequent transformation of the white-eye structures to honeycomb patterns. The arrows indicate the direction of particle motion in a single patch of the pattern. Temporal sequence after preparing the system: A) 9 min, B) 9.5 min, C)10 min, D) 13 min, E) 17 min, F) 24 min, G) 26 min, H) 30 min.*

the PA-MBO system is uncertain. Here, we use them as convenient tracers that reveal insights into the mechanical and/or hydrodynamic changes in the system. In the following experiments, each particle is resolved by approximately 5×5 pixels. Its movement is followed by computational particle tracking.

Figure 4 shows snapshots of a selected small area taken at different stages, starting from spatial homogeneity and leading via hexagons and white-eyes to honeycomb structures. During the early phases of the reaction, the particles perform a collective motion within the still liquid, layer (Figure 4A) and we could not detect any indication for macroscopic concentration patterns. This behavior ceases in the course of the reaction and gives way to a more organized motion illustrated in Figure 4B,C as well as in Figure 5A. The underlying data for this second phase is acquired 10 min after initiation of the reaction and, at this time, the system has reached a highly viscous, possibly gel-like state. Simultaneously, hexagonal patterns emerge that reflect the spatial patterning in the concentration of MB^+ described above. The particles are now moving towards the center of the bluish patches, one of which is captured in the image of Figure 5A (particle speed, 5-10 μm/s). This motion lasts for 1-2 min after which all particles cease to move. At this time visual inspection yields the first clear indications for surface deformations of the reaction layer.

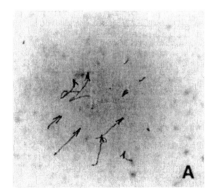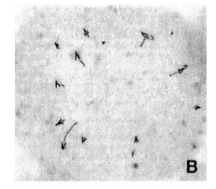

Figure 5. Measured particle traces superimposed onto the optically detected concentration pattern. Arrows mark the direction of particle movement during the early phase of the pattern formation (A), and during the transition to honeycomb patterns (B). Field of view: 1.7 × 1.6 mm². (Adapted with permission from reference 13. Copyright 2002 Oldenbourg Wissenschaftsverlag.)

The third, distinct phase of particle motion seems to be associated with secondary changes in the concentration pattern. The transition of hexagons to more complex structures, such as the "white-eye", honeycomb or chevron-like patterns is shown in Figures 2B-D, respectively (*12*). The transition from hexagons to honeycombs occurs through an intermediate "white-eye" state (Figure 2B and 4E) that slowly evolves from the initial pattern in a "one-patch-to-one-eye" fashion. During this continuous change no significant motion of the particles is detected. Subsequently, the "white-eye" patches undergo a transition in which the blue rings of the white eyes grow in size (cf. frames F and G of Figure 4) and give rise to the honeycomb-like pattern shown in Figure 2C (and also Figure 4H). This transition propagates rapidly through the entire system and often nucleates at defects of the original pattern. The front of this propagating transition can change the location of individual concentration patches and thereby reduce the number of defects (for details, see Ref. 11). It is during this phase that the position of the colloidal particles in the gel matrix is again affected. Figure 5B shows our measurements of the trajectories of numerous particles that now move from the center of the concentration patch outwards and, thus, retrace the motion observed during the second phase (Figure 5A).

After this outward motion is completed, all particles essentially remain immobilized. The macroscopic pattern of MB^+ concentration shows no significant changes any more except for a gradual fading of the bluish color. The surface deformations generated during the second and third phases of particle reorientation are permanent and withstand repetitive drying and swelling with water. Profilometric analyses of the patterned surface of a dried PA-MBO gel yielded a height difference of 3 μm between the indentations, the locations of which corresponded to the patches of the original concentration pattern.

Conclusions

A remarkable variety of semi-stationary concentration structures have been shown to form in the PA-MBO system. Irregular stripes and hexagons are the simplest and apparently fundamental patterns. Optical detection reveals that hexagons form at a time that is not easily discriminated from the gelation time of the acrylamide gel. Our data yield clear evidence for hydrodynamic flows during the early phases of the reaction (Figure 4A). Additional information is provided by our measurements of the pattern wavelength versus layer height (Figure 3B). The results suggest a proportional dependence and show that there is a critical layer height below which no concentration patterns are formed. Both findings appear to be in good agreement with the assumption of a surface-tension driven Marangoni instability. Although it is, hence, tempting to conclude that the patterns result from fluid flow, it remains puzzling that the hydrodynamic motion observed lacks spatial patterning and shows an insufficient temporal stability. Moreover, the height dependence of the pattern wavelength could also be caused by the influx of oxygen from the ambient atmosphere into the reaction layer. The underlying reaction-diffusion processes have to be taken into account, because they control the effective oxygen concentration and therefore the polymerization as well as the MBO reaction as a function of the layer height (*17*).

The first, *spatially organized* motion of the tracer particles correlates with the evolving concentration patches of the hexagonal pattern. Up to now, we are not able to answer the question whether this motion is driven by hydrodynamic flows in the highly viscous pre-gel solution or by mechanical deformations of the gelled medium. However, it is should be emphasized that at this stage the particle motion is always directed towards the center of the evolving patches of high MB^+ concentration. This finding is remarkable because hydrodynamic convection cells are expected to involve inward as well as outward flow of the fluid. Certainly, a precise determination of the gel point during pattern evolution is desirable. Since the gel point of the PA-MBO system depends on the intensity of illumination, its measurement using the methods proposed in (*18,19*) is especially difficult and further efforts in this direction are under way.

Mechanical instabilities of polymer systems usually arise in shrinking or swelling gels. A classic example is the formation of patterns on the surface of ionized acrylamide gels that swell due to an osmotically driven inflow of water (*20-22*). This inflow leads to mechanical stress within the gel and shear bending of the solvent-exposed gel layer. Despite the complex reactivity of the PA-MBO system, this type of mechanical instability seems to persist in our experiments and provides a possible explanation for the sudden, often front-like transition to honeycomb patterns. In this process, the reduction of mechanical stress is affected by the spatially periodic concentration patterns. Consequently, one can interpret the outward displacement of particles during the third phase of motion as a truly chemo-mechanical phenomenon.

Acknowledgments.

The author thanks O. Steinbock, E. Kasper and K. Tsuji for fruitful discussions and E. Slamova for providing the photograph in Fig. 1.

References

1. Kapral, R.; Showalter, K. (Eds.), *Chemical Waves and Patterns*, Kluwer Academic Publishers, Dordrecht, 1995.
2. Castets, V.; Dulos, E.; Boissonade, J.; DeKepper, P. *Phys. Rev. Lett.* **1990**, *64*, 2953-2956.
3. Quyang, Q.; Swinney, H. L. *Nature* **1991**, *352*, 610-611.
4. Hamik, C. T.; Manz, N.; Steinbock, O. *J. Phys. Chem.* **2001**, *A105*, 6144-6153.

5. Vanag, V. K.; Epstein, I. R. *Science* **2001**, *294*, 835-837.
6. Murray, J. D. *Mathematical Biology;* Springer-Verlag: Berlin-Heidelberg-New York, 1993.
7. Yoshida, R.; Takahashi, T.; Yamaguchi, T.; Ichijo, H. *J. Am. Chem. Soc.* **1996**, *118*, 5134-5135.
8. Watzl, M.; Münster, A. F. *Chem. Phys. Lett.* **1995**, *242*, 273-278.
9. Burger, M.; Field, R. J. *Nature* **1984**, *307*, 720-721.
10. Resch, P.; Münster, A. F.; Schneider, F. W. *J. Phys. Chem.* **1991**, *95*, 6270-6275.
11. Steinbock, O.; Kasper, E.; Müller, S. C. *J. Phys. Chem.* **1999**, *A103*, 3442-3446.
12. Müller, S. C.; Kasper, E.; Steinbock, O. *Macromol. Symp.* **2000**, *160*, 143-149.
13. Steinbock, O.; Kasper, E.; Müller, S. C. *Z. Phys. Chem.* **2002**, *216*, 687-697.
14. Watzl, M.; Münster, A. F. *J. Phys. Chem.* **1998**, *A102*, 2540-2546.
15. Fecher, F.; Strasser, P.; Eiswirth, M.; Schneider, F. W.; Münster, A. F. *Chem. Phys. Lett.* **1999**, *313*, 205-210.
16. Kurin-Csörgei, K.; Orban, M.; Zhabotinsky, A. M.; Epstein, I. R. *Chem. Phys. Lett.* **1998**, *295*, 70-74.
17. Münster, A. F. *Chem. Phys. Lett.* **1999**, *311*, 102-104.
18. Parreno, J.; Piérola, I. F. *Polymer* **1990**, *31*, 1768.
19. Lyubimova, T.; Caglio, S.; Gelfi, C.; Ringhetti, P. G.; Rabilloud, T. *Electrophoresis* **1993**, *14*, 40-50.
20. Tanaka, T.; Sun, S.-T.; Hirokawa, Y.; Katayama, S.; Kucera, J.; Hirose, Y.; Amiya, T. *Nature* **1987**, *325*, 796-798.
21. T. Hwa, T.; Kardar, M. *Phys. Rev. Lett.* **1988**, *61*, 106-109.
22. Suzuki, A.; Yoshikawa, S.; Bai, G. *J. Chem. Phys.* **1999**, *111*, 360-367.

Frontal Polymerization

Chapter 9

Nonlinear Dynamics in Frontal Polymerization

John A. Pojman[1], Sammy Popwell[1], Dionne I. Fortenberry[2], Vitaly A. Volpert[3], and Vladimir A. Volpert[4]

[1]Department of Chemistry and Biochemistry, University of Southern Mississippi, Hattiesburg, MS 39406
[2]Mississippi University for Women, Columbus, MS 39701
[3]MAPLY UMR 5585, Université Lyon I, F–69622 Villeurbanne Cedex, France
[4]Department of Engineering Sciences and Applied Mathematics, Northwestern University, Evanston, IL 60208–3125

Thermal frontal polymerization is a mode of converting monomer into polymer via a localized exothermic reaction zone that propagates through the coupling of thermal diffusion and the Arrhenius reaction kinetics of an exothermic polymerization. We review the range of nonlinear phenomena that have been observed in frontal polymerization systems and report new results on the role of gravity in spin modes and the development of spherically-propagating fronts.

Introduction

Frontal polymerization (FP) is a process of converting monomer into polymer via a localized reaction zone that propagates through the monomer. There are two modes of FP. Isothermal Frontal Polymerization (IFP), also called Interfacial Gel Polymerization, is a slow process in which polymerization occurs at a constant temperature and a localized reaction zone propagates because of the gel effect (*1,2*).

Thermal frontal polymerization involves the coupling of thermal diffusion and Arrhenius reaction kinetics of an exothermic polymerization. Frontal polymerization was discovered in Russia by Chechilo and Enikolopyan in 1972 (3). The literature up to 1984 was reviewed by Davtyan et al. (4). Pojman and his co-workers demonstrated the feasibility of traveling fronts in solutions of thermal free-radical initiators in a variety of neat monomers at ambient pressure using liquid monomers with high boiling points (5-7) and with a solid monomer, acrylamide (8,9). Fronts in solution have also been developed (10). The macrokinetics and dynamics of frontal polymerization have been examined in detail (11). A patented process has been developed for producing functionally-gradient materials (12,13).

Types of systems

The requirements for frontal polymerization are a system that does not react at the chosen initial temperature, but does react rapidly at an elevated temperature. The reaction must be exothermic.

Free-radical chemistry is the most widely used but not the only one. Frontal curing of epoxy resins has been studied (14-18). Begishev et al. studied anionic polymerization of ε-caprolactam (19). Frontal Ring-Opening Metathesis Polymerization (FROMP) has been successfully achieved with dicyclopentadiene (20) and applied to making IPNs (21). Mariani et al. have achieved FP with urethane chemistry (22).

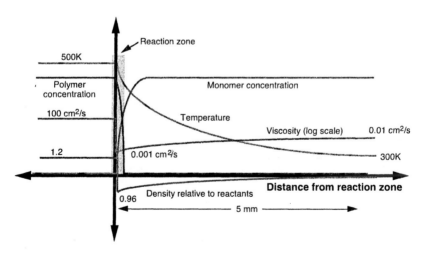

Figure 1. Schematic diagram showing changes in properties across a propagating polymerization front. (Courtesy of Paul Ronney.)

Frontal free-radical polymerization is fairly well understood. Studies on the velocity dependence on temperature and initiator concentration have been performed (7,11,23). Frontal polymerization in solution was performed (10), and initiators that do not produce gas were developed (24). The velocity can be affected by the initiator type and concentration but is on the order of a cm/min for monofunctional acrylates and as high as 20 cm/min for multifunctional acrylates (24).

We can also classify the systems based on the states of the reactants and products. Monofunctional monomers such as benzyl acrylate are liquids and produce liquid polymers, and we call these 'liquid/liquid' systems. Figure 1 shows a schematic of the changes in properties across a liquid/liquid front. Multifunctional monomers such as 1, 6 hexanediol diacrylate (HDDA) are liquid but produce a thermoset, solid product, and these we call 'liquid/solid' systems. Finally, solid monomers such as acrylamide (8,25) and transition metal nitrate complexes of acrylamide (26-28) can be polymerized frontally in 'solid/solid' systems.

Instabilities

Convective Instabilities

Because of the large thermal and concentration gradients, polymerization fronts are highly susceptible to buoyancy-induced convection. Garbey et al. performed the linear stability analysis for the liquid/liquid and liquid/solid cases (29-31). The bifurcation parameter was a 'frontal Rayleigh number':

$$R = g\beta q\kappa^2 / \nu c^3 \tag{1}$$

where g is the gravitational acceleration, β the thermal expansion coefficient, q the temperature increase at the front, κ the thermal diffusivity, ν the kinematic viscosity and c the front velocity.

Let us first consider the liquid/solid case. Neglecting heat loss, the descending front is always stable because it corresponds to heating a fluid from above. The front is always flat. If the front is ascending, convection may occur depending on the parameters of the system.

Bowden et al. experimentally confirmed that the first mode is an antisymmetric one, followed by an axisymmetric one (32). Figure 2 shows a flat descending front as well as axisymmetric and antisymmetric modes of ascending fronts. Figure 3 shows the stability diagram in the viscosity-front velocity plane. Most importantly, they confirmed that the stability of the fluid was a function not only of the viscosity but also of the front velocity. This means that the front dynamics affects the fluids dynamics, unlike with pH fronts and chemical waves in the BZ reaction in which the front velocity does not play a role in the stability of the fluid (33-38).

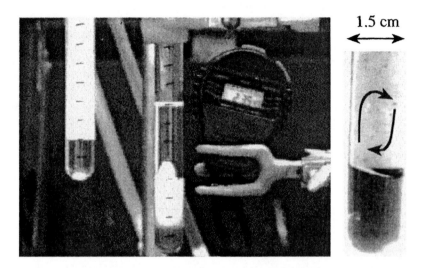

Figure 2. Left: The front on the left is descending and the one on the right ascending with an axisymmetric mode of convection. Right: An antisymmetric mode of an ascending front. The system is the acrylamid/bis-acrylamide polymerization in DMSO with persulfate initiator.

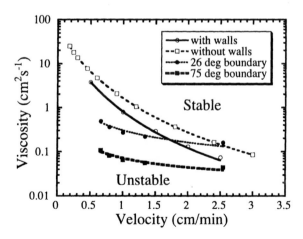

Figure 3. The stablity diagram for the system in Figure 2. Adapted from Bowden et al.

If the reactor is not vertical, there is no longer a question of stability – there is always convection. Bazile et al. studied descending fronts of acrylamide/bisacrylamide polymerization in DMSO as a function of tube orientation *(39)*. The fronts remained nearly perpendicular to the vertical but the velocity projected along the axis of the tube increased with a 1/cos θ dependence.

Liquid/liquid systems are more complicated than the previous case because a descending front can exhibit the Rayleigh-Taylor instability. Consider the schematic in Figure 1. The product is hotter than the reactant but is more dense, and because the product is a liquid, fingering can occur. Such front degeneration is shown in Figure 4. The Rayleigh-Taylor instability can be overcome using high pressure *(3)*, adding a filler *(23)*, using a dispersion in salt water *(40)* or performing the fronts in weightlessness *(40)*.

McCaughey et al. tested the analysis of Garbey et al. and found the same bifurcation sequence of antisymmetric to axisymmetric convection in ascending fronts *(41)* as seen with the liquid/solid case.

Garbey et al. also predicted that for a descending liquid/liquid front, an instability could arise even though the configuration would be stable for unreactive fluids *(29-31)*. This prediction has yet to be experimentally verified because liquid/liquid frontal polymerization exhibits the Rayleigh-Taylor instability. A thermal frontal system with a product that is less dense than the reactant is required.

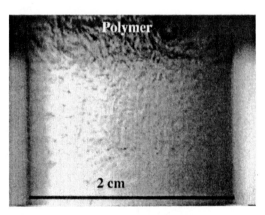

Figure 4. Rayleigh-Taylor instability with a descending front of butyl acrylate polymerization.

Texier-Picard et al. analyzed a polymerization front in which the molten polymer was immiscible with the monomer and predicted that a front could exhibit the Marangoni instability even though the comparable unreactive fluids would not exhibit the instability *(42)*. However, no liquid/liquid frontal system with an immiscible product has been identified. Even if such a system could be found, the experiment would have to be performed in weightlessness to prevent buoyancy-induced convection from interfering.

We note a significant difference between the liquid/liquid and the liquid/solid cases. For the liquid/solid case, convection in ascending fronts increases the front velocity but in the liquid/liquid case, convection slows the front. Convection increases the velocity of pH fronts and BZ waves. Why the difference between liquid/liquid frontal polymerization and other frontal systems? In liquid/liquid systems the convection also mixes cold monomer into the reaction zone, which lowers the front temperature. The front velocity depends more strongly on the front temperature than on the effective transport coefficient of the autocatalyst. Convection cannot mix monomer into the reaction zone of a front with a solid product but only increases thermal transport so the velocity is increased.

Thermal Instabilities

Fronts do not have to propagate as simple planar fronts. Analogously to oscillating reactions, a steady state can lose its stability as a parameter is varied and exhibit periodic behavior, either as pulsations or 'spin modes' in which a hot spot propagates around the reactor as the front propagates, leaving a helical pattern. This mode was first observed in Self-propagating High temperature Synthesis (SHS) (*43*).

The linear stability analysis of the longitudinally propagating fronts in the cylindrical adiabatic reactors with one overall reaction predicted that the expected frontal mode for the given reactive medium and diameter of reactor is governed by the Zeldovich number:

$$Z = \frac{T_m - T_o}{T_m} \frac{E_{eff}}{RT_m} \tag{2}$$

For FP, lowering the initial temperature (T_0), increasing the front temperature (T_m), increasing the energy of activation (E_{eff}) all increase the Zeldovich number. The planar mode is stable if $Z < Z_{cr} = 8.4$, and unstable if $Z > Z_{cr}$. By varying the Zeldovich number beyond the stability threshold, subsequent bifurcations leading to higher spin mode instabilities can be observed. Secondly, for a cylindrical geometry the number of spin heads or hot spots is also a function of the tube diameter. We point out that polymerization is not a one-step reaction, so that the above form of the Zeldovich number does not directly apply. However, estimates of the effective Zeldovich number can be obtained from estimates with the steady-state assumption for free-radical polymerization.

The most commonly observed case with frontal polymerization is the spin mode in which a 'hot spot' propagates around the front. A helical pattern is often observed in the sample. The first case was with the frontal polymerization of ε-caprolactam (*19,44*), and the next case was discovered by Pojman et al. in the methacrylic acid system in which the initial temperature was lowered (*45*).

Spin modes have also been observed in the frontal polymerization of transitional metal nitrate acrylamide complexes (*26,28*) but have not been observed in the frontal acrylamide polymerization system (*9*).

The single-head spin mode was studied in detail by Ilyashenko and Pojman (*46*). They were able to estimate the Zeldovich number using kinetic parameters for the initiator and the methacrylic acid. The value at room temperature was about 7, less than the critical value for spin modes. In fact, fronts at room temperature were planar and spin modes only appeared by lowering the initial temperature. However, spin modes could be observed by increasing the heat loss from the reactor by immersing the tube in water or oil. The simple analysis assumes an adiabatic system.

Effect of complex kinetics

Solovyov et al. performed a two-dimensional numerical study using a standard three-step free-radical mechanism (*47*). They calculated the Zeldovich number from the overall activation energy using the steady-state theory and determined the critical values for bifurcations to periodic modes and found that the complex kinetics stabilized the front.

Shult and Volpert performed the linear stability analysis for the same model and confirmed this result (*48*). Spade and Volpert studied linear stability for nonadiabatic systems (*49*). Gross and Volpert performed a nonlinear stability for the one-dimensional case (*50*). Commissiong et al. extended the nonlinear analysis to two dimensions (this volume). In the former analysis, they confirmed that, unlike in SHS (*51*), uniform pulsations are difficult to observe in frontal polymerization. In fact, no such one-dimensional pulsating modes have been observed.

An interesting problem arises in the study of fronts with multifunctional acrylates. At room temperature, acrylate like 1, 6 hexanediol diacrylate (HDDA) and triethylene glycol dimethacrylate (TGDMA) exhibit spin modes. In fact, if an inert diluent, such as dimethyl sulfoxide (DMSO) is added, the spins modes are more apparent even though the front temperature is reduced. Masere and Pojman found spin modes in the frontal polymerization of a diacrylate at ambient condition (*52*). Thus, although the mechanical quality of the resultant polymer material can be improved by using multifunctional acrylates, spin modes may appear and a nonuniform product results. This observation implicates the role of polymer crosslinking in front dynamics. In that same work, Masere and Pojman showed that pH indicators could be added to act as dyes that were bleached by the free-radicals, making the observation of the spin pattern readily apparent.

Tryson and Schultz studied the energy of activation of photopolymerized multifunctional acrylates and found it increased with increasing conversion because of crosslinking, which affects the propagation and termination steps (*53*). Gray found that the energy of activation of HDDA increased exponentially during the reaction (*54*). Using the steady-state theory of polymerization in tandem with Gray's results, Masere et al. calculated the effective energy of activation for thermally-initiated polymerization (photoinitiation has no energy of activation) by including the energy of activation of a typical peroxide (*55*). The Ea of HDDA polymerization increased from 80 kJ/mole at 0% conversion, the same as methacrylic acid, to a 140 kJ/mole at 80% conversion. This can explain how spin modes appear at room temperature with diacrylates but not monoacrylates. The Zeldovich number of methacrylic acid polymerization at room temperature is 7.2, below the stability threshold. Using the activation energy at the highest conversion that can be obtained with HDDA, Masere et al. estimated a Zeldovich number of 12 (*55*).

Masere et al. studied fronts with peroxide initiator at room temperature and used two bifurcation parameters (*55*). They added an inert diluent, DMSO, to change the front temperature and observed a variety of modes. More interestingly, they also varied the ratio of a monoacrylate, benzyl acrylate, to HDDA, keeping the front temperature constant.

The three-dimensional nature of the helixal pattern was studied by Manz et al. using Magnetic Resonance Imaging (MRI) (*56*). Figure 5 shows a reconstructed helix for a sample that was soaked with water.

Pojman et al. observed zigzag modes in square reactors and bistability in conical reactors (*57*).

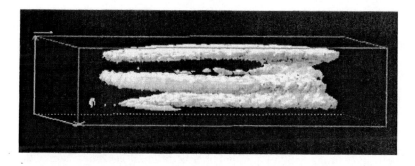

Figure 5. Reconstructed iso-surface from a T_2-weighted 3-dimensional image of a single-head spin mode (T_E = 40 ms). The voxel resolution is 0.12 mm in each direction. Sample was 1.5 cm in diameter and consisted of 1,6 hexanediol diacrylate in dimethyl sulfoxide with ammonium persulfate initiator. Image courtesy of B. Manz and F. Volke.

Effect of bubbles

Pojman et al. found an unusual mode of propagation when there are large amounts of bubbles (*45*). In studying fronts of methacrylic acid polymerization, they observed convection that periodically occurred under the front at the same time as the front deformed and undulated. The period of convection was about 20 seconds and remained constant during the entire front propagation.

Volpert et al. have analyzed the effect of the thermal expansion of the monomer on the thermal stability and concluded that the reaction front becomes less stable than without thermal expansion (*58*). The effective thermal expansion can be increased because of the bubbles, and it can considerably affect the stability conditions.

Experiments on bubbles in fronts were performed in weightlessness (*59*).

Effect of Convection

The physical mechanism leading to appearance of the spinning modes for polymerization fronts is the same as for combustion fronts. However, if the polymer or the monomer is in the liquid phase, then the properties of these regimes and the critical values of parameters when they appear can be influenced by hydrodynamics.

Consider a polymerization front with a liquid monomer and a solid polymer propagating upwards. If the Zeldovich number is sufficiently large, then the planar reaction front loses its stability resulting in appearance of a spinning mode. The high-temperature spots formed near the reaction front can lead to a convective motion of the liquid monomer above the front. This motion will mix the hot monomer near the front with the cold monomer above the front and, consequently, will decrease the temperature at the high-temperature spot. Therefore convection acts against the thermal instability, and the critical value of the Zeldovich number increases with increase of the Rayleigh number (*29*). In other words, there are two bifurcation parameters needed to describe the appearance of spin modes – the Zeldovich number and the frontal Rayleigh number, eq. 1.

If the front propagates downwards, the convective motion decreases the heat loss from the high-temperature spot to the unreacted monomer. The heat is conserved near the reaction front, and the perturbation of the temperature has better conditions to increase. Hence the critical value of the Zeldovich number decreases. We note that the influence of convection on the spinning modes for polymerization fronts with a liquid polymer is different in comparison with the case of a solid polymer (*31*).

The first experimental confirmation that gravity plays a role in spin modes in a liquid/solid system came in the study of descending fronts in which the viscosity was significantly increased with silica gel (*55*). Masere et al. found that silica gel significantly altered the spin behavior. Pojman et al. made a similar observation in square reactors (*57*).

The question arises why the analysis of Ilyashenko and Pojman worked so well for the methacrylic acid system, even though they did not consider the effect of convection. They induced spin modes by reducing the initial temperature to 0 °C – below the melting point of methacrylic acid. Thus, the system was now a solid/solid system and so hydrodynamics played only a small role in the melt zone.

We studied the dependence of spin modes on viscosity in the HDDA/persulfate system. Determining the viscosity is complicated by the shear-thinning behavior of silica gel suspensions. Figure 6 shows the apparent viscosity vs. shear rate for different percentages of silica gel in HDDA. The linear stability analysis assumes Newtonian behavior so we need to estimate the viscosity at zero shear, which is something we can not reliably estimate with our viscometer. We used the viscosities at the lowest shear rate we could measure and recognize that we are underestimating the true value. Fortunately, this does not affect the qualitative trends.

We measured the pitch of the helix by counting the number of spins over a 1 cm region of the sample. Figure 6 shows the spin behavior as a function of the apparent viscosity. Notice that the number of spins is independent of the viscosity until somewhere between 120 and 200 mPa s, when the spins disappear.

Three-dimensional Frontal Polymerization.

We can create spherically-propagating fronts using liquid/solid systems. By adding silica gel we can suppress convection but still have an initially transparent system. Figure 7 shows such a system in which a drop of photoinitiator was injected in the center of the HDDA/DMSO/persulfate solution that contained enough silica gel to make it highly viscous. By exposing the system to 365 nm radiation, the front ignites in the center.

It will be quite interesting to see how periodic modes manifest themselves in such a configuration.

Experimental

Viscosity of the fumed silica, Aerosil, (from Degussa) HDDA suspension was measured on a Brookfield viscometer. The percentages are weight/volume. The Aliquat persulfate was prepared according to the procedures in (*24*). 1, 6 hexanediol diacrylate (HDDA) (90% technical grade) was used as received from Aldrich.

All reactions were performed in a fume hood beyond in the unlikely event of a reactor bursting. Use of persulfate initiators makes this extremely rare.

116

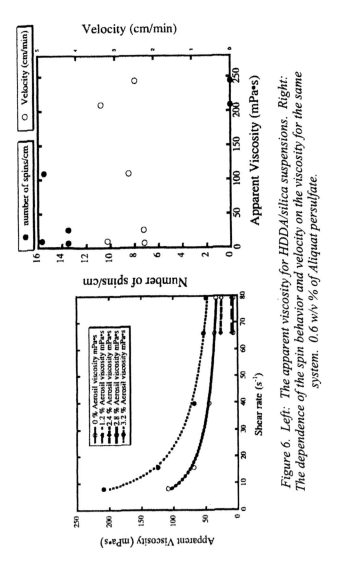

Figure 6. Left: The apparent viscosity for HDDA/silica suspensions. Right: The dependence of the spin behavior and velocity on the viscosity for the same system. 0.6 w/v % of Aliquat persulfate.

*Figure 7. A front propagating in a solution of 1,6 hexanediol diacrylate (40%)
and DMSO (60%) with ultrafine silica gel, 6.7% w/v. Ammonium persulfate
was the thermal initiator and Irgacure 184 the photoinitiator. Courtesy of
Marcus Molden.*

Conclusions

Thermal frontal polymerization exhibits the full range of nonlinear
dynamics phenomena, including those driven by hydrodynamics as well as
driven by intrinsic feedbacks in the chemistry. Features unique to
polymerization kinetics and properties allow the study of convection in fronts
and novel spin modes.

Acknowledgements

We acknowledge support from the NASA Reduced Gravity Materials
Science Program (NAG 8-973 and NAG 8-1858). and the National Science
Foundation (Grants CTS-0138660, CTS-0138712 and DMS-0103856).

References

1. Koike, Y.; Takezawa, Y.; Ohtsuka, Y. *Appl. Opt.* **1988**, *27*, 486-491.
2. Smirnov, B. R. M. k., S. S.; Lusinov, I. A.; Sidorenko, A. A.;Stegno, E. V.;
    Ivanov, V. V. *Vysokomol. Soedin., Ser. B* **1993**, *35*, 161-162.
3. Chechilo, N. M.; Khvilivitskii, R. J.; Enikolopyan, N. S. *Dokl. Akad. Nauk
    SSSR* **1972**, *204*, 1180-1181.

118

4. Davtyan, S. P.; Zhirkov, P. V.; Vol'fson, S. A. *Russ. Chem. Rev.* **1984**, *53*, 150-163.
5. Pojman, J. A. *J. Am. Chem. Soc.* **1991**, *113*, 6284-6286.
6. Pojman, J. A.; Craven, R.; Khan, A.; West, W. *J. Phys. Chem.* **1992**, *96*, 7466-7472.
7. Pojman, J. A.; Willis, J.; Fortenberry, D.; Ilyashenko, V.; Khan, A. *J. Polym. Sci. Part A: Polym Chem.* **1995**, *33*, 643-652.
8. Pojman, J. A.; Nagy, I. P.; Salter, C. *J. Am. Chem. Soc.* **1993**, *115*, 11044-11045.
9. Fortenberry, D. I.; Pojman, J. A. *J. Polym. Sci. Part A: Polym Chem.* **2000**, *38*, 1129-1135.
10. Pojman, J. A.; Curtis, G.; Ilyashenko, V. M. *J. Am. Chem. Soc.* **1996**, *118*, 3783-3784.
11. Pojman, J. A.; Ilyashenko, V. M.; Khan, A. M. *J. Chem. Soc. Faraday Trans.* **1996**, *92*, 2825-2837.
12. Pojman, J. A.; McCardle, T. W. U.S. Patent 6,057,406,2000
13. Pojman, J. A.; McCardle, T. W. U.S. Patent 6,313,237,2001
14. Surkov, N. F.; Davtyan, S. P.; Rozenberg, B. A.; Enikolopyan, N. S. *Dokl. Phys. Chem.* **1976**, *228*, 435-438.
15. Arutiunian, K. A.; Davtyan, S. P.; Rozenberg, B. A.; Enikolopyan, N. S. *Dokl. Akad. Nauk SSSR* **1975**, *223*, 657-660.
16. White, S. R.; Kim, C. Thermochemical modeling of an economical manufacturing technique for composite structures, *37th Int. SAMPE Symp. Exhib.,* **1992**, 240-251.
17. Davtyan, S. P.; Arutyunyan, K. A.; Shkadinskii, K. G.; Rozenberg, B. A.; Yenikolopyan, N. S. *Polymer Science U.S.S.R.* **1978**, *19*, 3149-3154.
18. Chekanov, Y.; Arrington, D.; Brust, G.; Pojman, J. A. *J. Appl. Polym. Sci.* **1997**, *66*, 1209-1216.
19. Begishev, V. P.; Volpert, V. A.; Davtyan, S. P.; Malkin, A. Y. *Dokl. Akad. Nauk SSSR* **1973**, *208*, 892.
20. Mariani, A.; Fiori, S.; Chekanov, Y.; Pojman, J. A. *Macromolecules* **2001**, *34*, 6539-6541.
21. Fiori, S.; Mariani, A.; Ricco, L.; Russo, S. *e-Polymers* **2002**, *29*, 1-10.
22. Fiori, S.; Mariani, A.; Ricco, L.; Russo, S. *Macromolecules* **2003**, *36*, 2674-2679.
23. Goldfeder, P. M.; Volpert, V. A.; Ilyashenko, V. M.; Khan, A. M.; Pojman, J. A.; Solovyov, S. E. *J. Phys. Chem. B* **1997**, *101*, 3474-3482.
24. Masere, J.; Chekanov, Y.; Warren, J. R.; Stewart, F.; Al-Kaysi, R.; Rasmussen, J. K.; Pojman, J. A. *J. Poly. Sci. Part A. Polym. Chem.* **2000**, *38*, 3984-3990.
25. Fortenberry, D.; Pojman, J. A. *Polym. Prepr. (Am Chem. Soc. Div. Polym. Chem.)* **1997**, *38(2)*, 472-473.
26. Savostyanov, V. S.; Kritskaya, D. A.; Ponomarev, A. N.; Pomogailo, A. D. *J. Poly. Sci. Part A: Poly. Chem.* **1994**, *32*, 1201-1212.
27. Dzhardimalieva, G. I.; Pomogailo, A. D.; Volpert, V. A. *J. Inorganic and Organometallic Polymers* **2002**, *12*, 1-21.

28. Barelko, V. V.; Pomogailo, A. D.; Dzhardimalieva, G. I.; Evstratova, S. I.; Rozenberg, A. S.; Uflyand, I. E. *Chaos* **1999**, *9*, 342-347.
29. Garbey, M.; Taik, A.; Volpert, V. *Quart. Appl. Math.* **1998**, *56*, 1-35.
30. Garbey, M.; Taik, A.; Volpert, V. *Preprint CNRS* **1994**, *187*, 1-42.
31. Garbey, M.; Taik, A.; Volpert, V. *Quart. Appl. Math.* **1996**, *54*, 225-247.
32. Bowden, G.; Garbey, M.; Ilyashenko, V. M.; Pojman, J. A.; Solovyov, S.; Taik, A.; Volpert, V. *J. Phys. Chem. B* **1997**, *101*, 678-686.
33. Pojman, J. A.; Epstein, I. R. *J. Phys. Chem.* **1990**, *94*, 4966-4972.
34. Pojman, J. A.; Epstein, I. R.; McManus, T.; Showalter, K. *J. Phys. Chem.* **1991**, *95*, 1299-1306.
35. Pojman, J. A.; Epstein, I. R.; Nagy, I. *J. Phys. Chem.* **1991**, *95*, 1306-1311.
36. Masere, J.; Vasquez, D. A.; Edwards, B. F.; Wilder, J. W.; Showalter, K. *J. Phys. Chem.* **1994**, *98*, 6505-6508.
37. Plesser, T.; Wilke, H.; Winters, K. H. *Chem. Phys. Lett.* **1992**, *200*, 158-162.
38. Wu, Y.; Vasquez, D. A.; Edwards, B. F.; Wilder, J. W. *Phys. Rev. E* **1995**, *51*, 1119-1127.
39. Bazile Jr., M.; Nichols, H. A.; Pojman, J. A.; Volpert, V. *J. Polym. Sci. Part A: Polym Chem.* **2002**, *40*, 3504-3508.
40. Pojman, J. A.; Gunn, G.; Owens, J.; Simmons, C. *J. Phys. Chem. Part B* **1998**, *102*, 3927-3929.
41. McCaughey, B.; Pojman, J. A.; Simmons, C.; Volpert, V. A. *Chaos* **1998**, *8*, 520-529.
42. Texier-Picard, R.; Pojman, J. A.; Volpert, V. A. *Chaos* **2000**, *10*, 224-230.
43. Maksimov, Y. M.; Pak, A. T.; Lavrenchuk, G. V.; Naiborodenko, Y. S.; Merzhanov, A. G. *Comb. Expl. Shock Waves* **1979**, *15*, 415-418.
44. Begishev, V. P.; Volpert, V. A.; Davtyan, S. P.; Malkin, A. Y. *Dokl. Phys. Chem.* **1985**, *279*, 1075-1077.
45. Pojman, J. A.; Ilyashenko, V. M.; Khan, A. M. *Physica D* **1995**, *84*, 260-268.
46. Ilyashenko, V. M.; Pojman, J. A. *Chaos* **1998**, *8*, 285-287.
47. Solovyov, S. E.; Ilyashenko, V. M.; Pojman, J. A. *Chaos* **1997**, *7*, 331-340.
48. Shult, D. A.; Volpert, V. A. *Int. J. of SHS* **1999**, *8*, 417-440.
49. Spade, C. A.; Volpert, V. A. *Combustion Theory and Modelling* **2001**, *5*, 21-39.
50. Gross, L. K.; Volpert, V. A. *Studies in Appl. Math.* **2003**, *110*, 351-376.
51. Shkadinsky, K. G.; Khaikin, B. I.; Merzhanov, A. G. *Combust. Explos. Shock Waves* **1971**, *1*, 15-22.
52. Masere, J.; Pojman, J. A. *J. Chem. Soc. Faraday Trans.* **1998**, *94*, 919-922.
53. Tryson, G. R.; Shultz, A. R. *J. Polym. Sci. Polym. Phys. Ed.* **1979**, *17*, 2059-2075.
54. Gray, K. N. Master's thesis, University of Southern Mississippi, Hattiesburg, MS, 1988.
55. Masere, J.; Stewart, F.; Meehan, T.; Pojman, J. A. *Chaos* **1999**, *9*, 315-322.
56. Manz, B.; Masere, J.; Pojman, J. A.; Volke, F. *J. Poly. Sci. Part A. Polym. Chem.* **2001**, *39*, 1075-1080.

57. Pojman, J. A.; Masere, J.; Petretto, E.; Rustici, M.; Huh, D.-S.; Kim, M. S.; Volpert, V. *Chaos* **2002**, *12*, 56-65.
58. Volpert, V. A.; Volpert, V. A.; Pojman, J. A. *Chem. Eng. Sci.* **1994**, *14*, 2385-2388.
59. Ainsworth, W.; Pojman, J. A.; Chekanov, Y.; Masere, J. Bubble Behavior in Frontal Polymerization: Results from KC-135 Parabolic Flights: in *Polymer Research in Microgravity: Polymerization and Processing, ACS Symposium Series No. 793*; Downey, J. P. Pojman, J. A., Ed.; American Chemical Society: Washington, DC, 2001; pp 112-125.

Chapter 10

Recent Chemical Advances in Frontal Polymerization

Alberto Mariani[1], Stefano Fiori[1], Giulio Malucelli[2], Silvia Pincin[3], Laura Ricco[3], and Saverio Russo[3]

[1]Dipartimento di Chimica, Università di Sassari, and Local INSTM Research Unit, Via Vienna 2, 07100 Sassari, Italy
[2]Dipartimento di Scienza dei Materiali ed Ingegneria Chimica, Politecnico di Torino, and Local INSTM Research Unit, Corso Duca Degli Abruzzi 24, 10129 Torino, Italy
[3]Dipartimento di Chimica e Chimica Industriale, Università di Genova, and Local INSTM Research Unit, Via Dodecaneso 31, 16146 Genova, Italy

Frontal Polymerization is a technique of polymer synthesis in which the heat released by the reaction generates a self-sustaining polymerization front able to propagate through a reactor, thus converting monomer into polymer. After the ignition and the subsequent front formation, no further energy supplying is requested for the polymerization to occur. The syntheses can be performed in bulk or in solution; solid monomers, if molten at the front temperature, can also be used. Furthermore, monomers can be successfully used without removing the inhibitor, if present. The polymers previously obtained by this technique were polyacrylates, nylons and epoxy resins. In this chapter we present a short overview of our recent developments in Frontal Polymerization that allow us to use this technique for the preparation of polyurethanes, polyester/styrene resins, polydicylopentadiene and its IPNs with polyacrylates. Eventually, we give some preliminary details on the successful application of FP to the consolidation of stones.

Introduction

The main chemical reactions having a practical interest are generally included in one of the following three classes. a) the reactions happening for simple reactant mixing at room temperature, or below; b) the reactions happening in heated reactors or molds (i.e., with a continuous energy providing); c) the reactions undergo by those systems which are inert at room temperature (or below) but which react fast, and without further energy providing, if ignited (e.g. combustion, explosions).

In the field of polymer synthesis, while the first two classes are widely used and studied, the third one is almost unstudied and it has not been used in practice so far. However, having suitable polymerizable systems belonging to class c) should be regarded as a desirable feature in many practical applications.

Sometimes, if the monomer system can be polymerized by the effect of light, it is possible to use photopolymerization instead of heating. However, such a technique, although very useful for curing thin samples, cannot be successfully used in those cases (that probably represent the majority) in which thick materials have to be cured, especially if they are opaque (a very common feature, especially in the case of composites). Indeed, it is well known that the resulting light intensity gradient prevents the possibility of photopolymerizing materials of thickness larger than a few centimeters.

At least in these cases, Frontal Polymerization (FP) might represent a possible valid alternative technique having practical and economical advantages. FP exploits the heat production due to exothermicity of the reaction itself and its dispersion by thermal conduction. If the amount of dissipated heat is not too large, a sufficient amount of energy able to induce the polymerization of the monomer close to the hot zone is provided. The result is the formation of a hot polymerization front capable of self-sustaining and propagating through the reactor.

Therefore, the lack of heat sources, generally requested for carrying out many traditional chemical reactions, results in many advantages in terms of cost and environmental impact. Furthermore, FP is generally (but not exclusively) applicable to solvent-free systems. Because of the high temperatures typically reached by the fronts, monomers can be used as received, i.e. without the usual elimination of the dissolved inhibitor. Moreover, high conversions are often found, thus making the final common purification procedures not necessary.

The first studies on FP were performed in the former USSR on methyl methacrylate that was polymerized under drastic pressure conditions (> 3000 atm) (*1, 2*). Lately, several other vinyl monomers have been polymerized at ambient pressure (*3-5*). Furthermore, FP was applied to the curing process of epoxy-based materials (*6-8*).

In some cases, front instabilities have been found and a number of methods for stabilizing fronts have been proposed (*4, 9,*). One of the most frequent instability is the so-called phenomenon of "fingering". This event occurs when a polymer is denser than the corresponding monomer and melts at the front temperature; the result is the formation of descending drops of hot polymer material that contaminate the lower zones of unreacted monomer, thus causing heat removal from the front (sometimes even quenching it) and local spontaneous polymerization (SP) ignition.

Actually, for many years the only chemical systems suitable for FP have been those mentioned above and much of the research in this field was devoted to the study of front instabilities (*9-16*) and to modelling. (*17, 18*).

In 2000, we have begun our research on this topic paying particular attention to the possibility of finding new chemical systems able to frontally polymerize and new FP applications. Specifically, we have studied polyurethanes (*19, 20*), polyester/styrene resins (*21*), polydicylopentadiene (*22*) and its IPNs with polyacrylates (*23*). Furthermore, we were able to prepare films (*24*) and to apply FP to the consolidation of porous materials (i.e. stones, woods, flaxes, papers), in particular –but not only- those having a historical-artistic interest (*24*). In this chapter we present a brief overview of these recent findings.

General Experimental Considerations

Usually, the experimental work on a new system potentially able to undergo FP is devoted to determine the conditions in which a *pure* Frontal Polymerization takes place. With this term we indicate an FP as the unique mode of polymerization occurring in the reactor at a given time. In particular, we mean that SP is not happening.

For pure FP to occur, the investigated chemical system must be almost inert at relatively low temperature but very reactive at the temperature reached by the front (generally much higher than 100 °C). That means that sufficiently long pot-lives are necessary. However, the polymerizing system does not always have this requisite and, sometimes, a proper additive has to be added.

Pure FP is often characterized by constant front velocity (V_f), but such a behavior is not always found. Sometimes, this is due to the bubbles formed during the process which impart a non-constant movement to the front.

Furthermore, a linear dependence between front position and time can be found also if SP is simultaneously occurring. For instance, that happens if FP is relatively fast if compared with SP in the time scale needed to complete the polymerization in a reactor of a certain geometry (the longer a reactor is, the more crucial this point becomes).

For the above reasons, temperature of the monomer far from the incoming front is a more significant parameter to be checked. Indeed, if a temperature increase happens in regions distant from the hot travelling zone, SP is probably taking place. A convenient method to verify the occurrence of that is given by the analysis of temperature profiles (Figure 1).

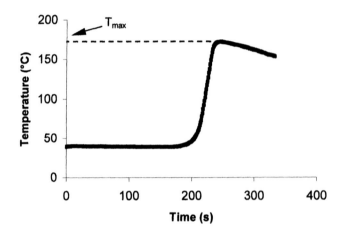

Figure 1. Temperature profile recorded during a typical pure FP experiment.

These plots provide many useful information, namely:

a)   the extent of temperature jump and the value of the maximum reached (T_{max}, see Figure 1): this latter is very important for the final characteristics of the polymer because it is related to the degrees of conversion and crosslinking, the onset of polymer degradation, bubble formation etc.;

b)   the range interested to heating: it is the monomer zone close to the incoming polymerization front; *e.g.*, its presence can ensure the FP of monomers which are solid at room temperature but that melt immediately before being crossed by the front (*5, 22*);

c)   the "degree of adiabaticity": the larger the heat loss is, the higher the slope of the curve after T_{max} is;

d) the presence of simultaneous SP: if no increments of temperature are recorded before the arrival of the front, SP is probably not occurring.

As mentioned above, sometimes SP is not due to the relatively short pot-life of the reacting system but it has to be attributed to the formation of fingering. This phenomenon can be reduced or prevented by increasing medium viscosity by adding suitable additives (4, 9, 13, 19, 20). If possible, they should be chosen among those imparting desirable features to the final polymer material (19, 20).

Frontal Polymerization of New Chemical Systems

Polyurethanes

The first polyurethane obtained by FP has been derived by the reaction between 1,6-hexamethylene diisocyanate (HDI) and ethylene glycol (EG) in the presence of dibutyltin dilaurate (DBTDL) as a catalyst, and pyrocatechol as an additive necessary for ensuring a sufficiently long pot-life (25). Indeed, if this latter compound is not present, instantaneous SP occurs. Furthermore, fingering was avoided by adding 3 wt.-% fumed silica (Cabosil®) to the above components dissolved in 18 wt.-% dimethylsulfoxide (DMSO).

A thorough investigation on the effect of all the above components have been performed and the ranges of their relative concentrations in which pure FP occurs have been determined (19, 20). On this respect, it was found that a [pyrocatechol]/[DBTDL] molar ratio ≥ 11 prevents SP, and the higher the pyrocatechol content is, the lower T_{max} and V_f are.

About the DBTDL effect, an increase of its amount causes a corresponding rise of both T_{max} and V_f. For brevity, only the dependence of T_{max} on the [DBTDL]/[HDI] molar ratio is shown in Figure 2. The experimental points indicated in this plot define the range in which pure FP has been observed.

Table I lists the [η] of some samples obtained by FP compared with that of the corresponding batch ones.

Samples F1-3 have been synthesized by using different amounts of DBTDL. As already mentioned, the larger the catalyst content is, the higher T_{max} is. Besides, it is noteworthy that intrinsic viscosity increases as well. Namely, FP3 sample is characterized by [η] = 0.76 dl/g, that is the larger value found in this work (19).

In order to compare the two techniques, we have performed a batch synthesis following the recipe given in literature (25) (80 °C for 1h) by using the same feed composition of FP3 (sample B2) obtaining a [η] value equal to 0.46 dl/g, much lower than those of the corresponding FP samples.

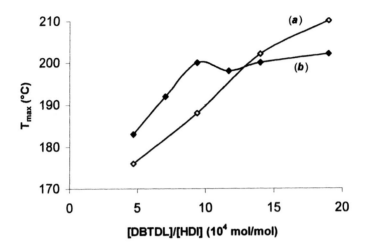

Figure 2. Dependence of T_{max} on the [DBTDL]/[HDI] molar ratio ((a):
[pyrocatechol]/[DBTDL] = 11 mol/mol; (b): [pyrocatechol]/[DBTDL] = 28
mol/mol). Adapted from Reference 19.

Table I. Intrinsic Viscosity Data of Some Polyurethane Samples Obtained by FP and Batch Polymerization

Sample	[DBTDL]/[HDI] (mol/mol)	T_{bath} [a] (°C)	T_{max} (°C)	$[\eta]$ [b] (dl/g)
B1 [c]	$1.4 \cdot 10^{-3}$	180	-	0.23
B2 [c]	$1.4 \cdot 10^{-3}$	80	-	0.46
FP2 [d]	$4.7 \cdot 10^{-4}$	-	176	0.20
FP1 [d]	$9.4 \cdot 10^{-4}$	-	188	0.47
FP3 [d]	$1.4 \cdot 10^{-3}$	-	202	0.76

NOTES: *a)* reaction temperature; *b)* in *m*-cresol at 25 °C; *c)* samples obtained by batch polymerization; *d)* samples obtained by FP. Other reaction conditions common to all samples: [pyrocatechol]/[DBTDL] = 11; DMSO = 18 wt.-%; Cabosil = 3 wt.-%.

SOURCE: Adapted from Reference 19.

The synthesis B1 was done to check if raising temperature could have a positive effect on the molecular weight. In particular, this experiment was carried out for 1 h at 180 °C, i.e., in the same range typically reached by the hot propagating fronts. As a matter of the fact, $[\eta] = 0.23$ dl/g was found, thus indicating that such a high temperature kept for long time results in a probable degradation of the polyurethane (we remember that, in the case of FP, the maximum temperature is reached in a few seconds and is immediately followed by cooling).

As a conclusion, it should be emphasized that by FP polyurethanes characterized by $[\eta]$ (i.e. molecular weight) higher than those prepared by the classical batch procedure are obtainable. Moreover, the reaction takes much shorter times and does not require external heating except for the ignition.

After this first example, other polyurethane systems have been successfully investigated, namely those based on tolylene-2,4-diisocyanate, 4,4'-methylenediphenylisocyanate, polycaprolactone diol, butene diol, butine diol. The related results will be reported in a next paper.

Unsaturated Polyester Resins

Unsaturated polyester resins (UPER) are largely used as components in polymer composites (*26, 27*) Generally, they are obtained by reaction of an unsaturated polyester (UPE) with a vinyl monomer. In our experiments, we have cured the UPE represented in Figure 3 with styrene (STY), which is the monomer mostly used for this purpose, or hydroxyethyl acrylate (HEA) and/or a diblocked diisocyanate (i.e., acrylic acid 2-[6-(2-acryloyloxy-ethoxycarbonylamino)-hexylcarbamoyloxy]-ethyl ester, UA). The reaction is highly exothermic, thus allowing for a rapid FP.

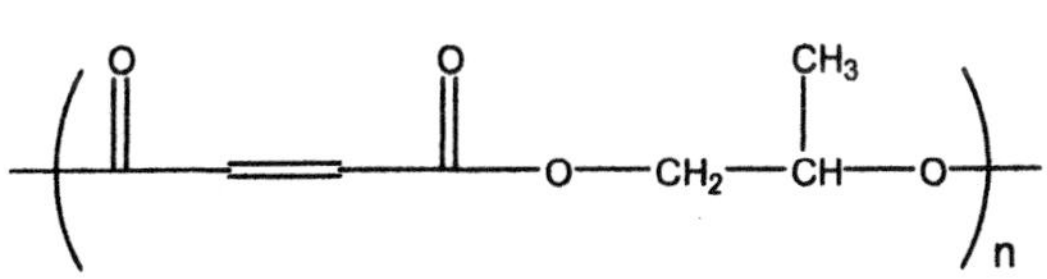

Figure 3. The unsaturated polyester used in our studies.

Also in these studies, the relative amounts of all components have been varied in order to find the conditions in which pure FP occurs. When STY was used as a curing agent, it was found that its amount can range between 20 and 40 wt.-%. Indeed, at lower content no FP has been observed even if large amounts of radical initiator (benzoyl peroxide, BPO) were used. Conversely, when more than 40 wt.-% STY was used, phase separation occurred.

Figure 4 shows the dependence of T_{max} on the type and the amount of radical initiator for a content of STY equal to 30 wt.-% (the amount typically present in commercial products). Three initiators, belonging to three different chemical classes, have been tested: BPO, azobisisobutyronitrile (AIBN) and Aliquat® persulfate (APS) (28).

An increment of T_{max} due to the increase of the amount of initiator was found in any set of experiments, an effect being particularly evident in the case of BPO and AIBN.

At variance, in the same range of relative concentrations (initiator = 1-10 mol %), the APS curve does not rise as much significantly.

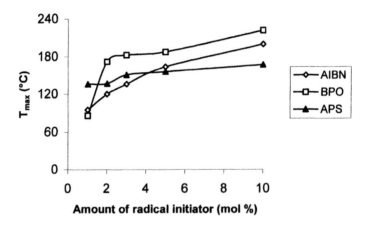

Figure 4. Dependence of T_{max} *on the type and amount of initiator (*w_s *= 30 wt.-%). Adapted from Reference 21.*

However, this result is probably due to the dilution effect that such a high molecular weight initiator exerts, thus dispersing heat and not allowing for large temperature increments. UPER materials obtained by FP have been compared with the corresponding samples prepared by the classical batch method. DSC analysis has shown that FP permits to reach quantitative conversion in a very short time, whereas by the batch technique much lower conversion has been obtained even though longer reaction times have been used. Moreover, the FP copolymer is characterized by higher T_g than the batch one (136 vs. 92 °C, first scan; 153 vs. 106 °C, second scan).

On the basis of what reported in Ref. 29, STY was replaced by HEA and by UA as curing agents. Those compounds have the advantages of being less volatile and impart better mechanical properties to the UPER than what STY does.

For instance, the following system was studied: UPE (70 wt.-%, kept invariable) + HEA + UA. The ratio between these two latter compounds has been varied by keeping constant the amount of APS as the radical initiator (3 mol % referred to the number of vinyl groups). By increasing the amount of UA, a slight raise of T_{max} was found: from 124 °C at UA/HEA = 0, to 147 °C at UA/HEA = 0.5 w/w, this latter being the maximum amount of UA which ensures a homogenous solution of all the reaction components. In the same range, V_f keeps constant at 0.6 cm/min.

At variance, this parameter was found to be slightly dependent on the amount of APS. Indeed, in the case of the mixture having this following weight composition: PE : HEA : UA = 70 : 21 : 9, V_f ranged between 0.4 (at APS = 1 mol %) to 0.5 cm/min (at APS= 10 mol %).

Polydicyclopentadiene and Related Materials

In 2001, in collaboration with Pojman's group we have published a work on the first example of Frontal Ring Opening Metathesis Polymerization (FROMP). Namely, dicyclopentadiene (DCPD) was polymerized by this technique.

Polymerization runs were performed by dissolving bis(tricyclohexylphosphine)benylidine ruthenium(IV) (Grubbs' catalyst, GC) into the monomer, already containing a proper amount of triphenylphosphine (PPh_3). This additive behaves as an inhibitor thus allowing for a longer pot-life. Moreover, as soon as the above mixtures were prepared, they were cooled to 27 °C. This means that FROMP experiments were performed on solid monomer mixtures which melt immediately before being reached by the hot polymerization fronts.

The effect of the relative ratios among DCPD, catalyst and PPh_3 on V_f and T_{max} were studied and the corresponding results have been published in Ref. 22.

Lately, our research on this polymer has been extended to its Interpenetrating Polymer Networks (IPNs) with poly(acrylate)s. These materials are used as components of structural composites in the automotive and electronic industries (e.g. in printed circuit boards). Among them, the most used is that made of poly(DCPD) and PMMA (30).

In our research work, we have explored the possibility of preparing IPNs made of poly(DCPD) and PMMA or poly(triethyleneglycol dimethacrylate) (poly(TGDMA)). While TGDMA is widely studied in FP, MMA is not able to undergo Frontal Polymerization at room pressure if alone. Consequently, one of the main goals of such a research was that of ascertaining if IPNs based on DCPD and MMA can be prepared by simultaneous non-interfering FPs exploiting the heat released during the DCPD FROMP in order to sustain also the MMA front.

The reaction mixtures were constituted by DCPD, GC, PPh₃, MMA and a suitable radical initiator (AIBN or BPO or APS).

Initially, it was determined that the amount of DCPD which ensures the above simultaneous FPs should be at least equal to 75 %.

As an example, the dependence of V_f on the GC content is listed in Table II, and that on the type and amount of radical initiator is shown in Figure 5.

Table II. Dependence of V_f on the GC Content in the Preparation of an IPN Made of DCPD and MMA

[DCPD]/[GC] (mol/mol)	V_f (cm/min)	T_{max} (°C)
750	1.6	171
1000	1.2	156
1500	1.1	133

NOTE: [PPh3] / [GC] = 3 mol/mol.
SOURCE: Adapted from Reference 23.

As predictable, when increasing amounts of BPO or AIBN were used, an increase of both V_f and T_{max} was found. On the contrary, the larger the amount of APS was, the smaller those parameters were. This apparent inconsistency can be explained by the observation that even little quantities of APS strongly depress GC activity and, in turn, that one of the whole polymerizing system. This observation being a confirmation that the general process is *ruled* by the DCPD FROMP which allows for the simultaneous frontal polymerization of MMA, otherwise not obtainable in similar pressure conditions.

Furthermore, it has been demonstrated that the number of monomers to be frontally polymerized could significantly be extended by including those not able to undergo FP if alone, but that could work in the presence of another chemical system (not necessarily a monomer) able to sustain their travelling front.

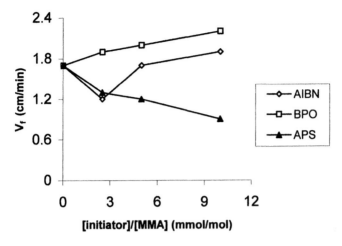

Figure 5. Dependence of V_f on the type and amount of radical initiator in the preparation of an IPN made of DCPD and MMA. Adapted from Reference 23.

Application of the Frontal Polymerization Technique to the Consolidation of Porous Materials

On the basis of what previously reported in literature (*31*) and after some preliminary studies carried out in our laboratories, it has been found that FP can work into porous objects such as textiles, woods, stones, (*24, 31*) papers, etc.

These materials are soaked with a monomer solution (already containing an initiator) and FP is ignited by heating a localized zone onto their surface. Namely, *Finale Stone* has been successfully tested finding that FP represents a very useful method of protection. Actually, by comparing the former common approach (i.e. the soaked sample is kept in an oven at 60 °C for a period sufficient for the polymerization to occur) with the FP technique, it has been found that this latter method gives better results. For instance, measurements of capillary water absorption have been done on some specimens treated following the above protocols and the data have been compared with that found by testing a non-treated sample. The results, reported in Figure 6, clearly indicate that FP permits to obtain the best protection. A paper dealing with these findings and containing the data concerning all tests is in preparation.

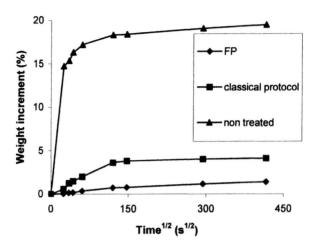

Figure 6. Weight increment due to capillary water absorption in an FP treated sample compared with a "classically" treated (see text) and a non-treated one.

Experimental Part

FP in Tubular Reactors

In a typical run, a glass test tube is loaded with the proper amounts of reaction components (momomer(s), initiator and/or catalyst, and -when necessary- additive(s) and/or solvent). FP is triggered by heating the upper end of the tube with a soldering iron tip. A thermocouple probe linked to a digital temperature reader is used for recording temperature profile (sampling rate: 1 Hz).

FP Inside Stones

In a typical run, a cubic stone specimen (5 x 5 x 5 cm) is soaked by capillary absorption with a monomer containing APS or AIBN or BPO. FP runs are started by heating the lower base of the sample with a heated plate.

133

Acknowledgment

The authors wish to thank S. Bidali, E. Pedemonte, J.A. Pojman, M. Sangermano, Ester Sanna and S. Vicini for their assistance in this work. MIUR-URST funds are gratefully acknowledged.

References

1. Chechilo, N.M.; Enikolopyan, N.S. *Dokl. Phys.Chem.* **1975**, *221*, 392.
2. Chechilo, N.M.; Enikolopyan, N.S. *Dokl. Phys.Chem.* **1976**, *230*, 840.
3. Pojman, J. A. *J. Am. Chem. Soc.* **1991**, *113*, 6284.
4. Khan, A. M.; Pojman, J. A. *Trends Polym. Sci.* **1996**, *4*, 253.
5. Fortenberry, D. I.; Pojman, J. A. *J. Polym. Sci., Part A: Polym Chem.* **2000**, *38*, 1129.
6. Chekanov, Y.; Arrington, D.; Brust, G.; Pojman, J.A. *J. Appl. Polym. Sci.* **1997**, *66*, 1209.
7. Kim, C.; Teng, H.; Tucker, C.L. III; White, S.R. *J. Comp. Mater.* **1995**, *29*(9), 1222.
8. Pojman, J.A.; Elcan, W.; Khan, A.M.; Mathias, L. *J. Polym. Sci., Part A: Polym. Chem.* **1997**, *35*, 227.
9. Pojman, J.A.; Ilyashenko, V.M.; Khan, A.K. *J. Chem. Soc., Faraday Trans.* **1996**, *92*(16), 2825.
10. McCaughey, B.; Pojman, J.A.; Simmons, C.; Volpert, V.A. *CHAOS* **1998**, *8*(2), 520.
11. Bowden, G.; Garbev, M.; Ilyashenko, V.M.; Pojman, J.A., Solovyov, S.E.; Taik, A.; Volpert, V.A. *J. Phys. Chem. B* **1997**, *101*(4), 678.
12. Ilyashenko, V.M.; Pojman, J.A. *CHAOS* **1998**, *8*(1), 285.
13. Pojman, J.A.; Craven, R.; Khan, A.; West, W. *J. Phys. Chem.* **1992**, *96*, 7466.
14. Nagy, I.P.; Pojman, J.A. *J. Phys. Chem.* **1996**, 100, 3299.
15. Masere, J.; Stewart, F.; Meehan, T.; Pojman, J.A. *CHAOS* **1999**, *9*(2), 315.
16. Epstein, I.R.; Pojman, J.A. *CHAOS* **1999**, *9*(2), 255.
17. Solovyov, S.E.; Ilyashenko, V.M.; Pojman, J.A. *CHAOS* **1997**, *7*(2), 331.
18. Goldfeder, P.M.; Volpert, V.A.; Ilyashenko, V.M.; Khan, A.M.; Pojman, J.A.; Solovyov, S.E. *J. Phys. Chem. B* **1997**, 101, 3474.
19. Fiori S.; Mariani A.; Ricco, L.; Russo, S. *Macromolecules* **2003**, *000*, 0000.
20. Mariani A.; Fiori S.; Ricco, L.; Russo, S. *ACS Polym. Prepr.* **2002**, *43(2)*, 875.
21. Fiori S.; Malucelli, G.; Mariani A.; Ricco, L.; Russo, S. *e-Polymers* **2002**, *057*, 1.

22. Mariani, A.; Fiori, S.; Chekanov, Y.; Pojman, J.A. *Macromolecules* **2001**, *34*, 6539.
23. Fiori S.; Mariani A.; Ricco, L.; Russo *e-Polymers* **2002**, *029*, 1.
24. Mariani A.; Fiori S.; Pedemonte, E.; Pincin, S.; Ricco, L.; Russo, S. *ACS Polym. Prepr.* **2002**, *43*(2), 814.
25. Dammann, L.G.; Carlson, G.M. U.S. Patent 4,788,083, **1988**.
26. *Comprehensive Polymer Science*; Eastmond, G.C., Ledwith, A.; Russo, S.; Sigwalt, P. Eds., Pergamon Press: Oxford, UK, 1989, Vol. 5, pp. 331-344.
27. Pusatcioglu, S.Y.; Fricke, A.L.; Hassler, J.C. *J. Appl. Polym. Sci.* **1979**, *24*, 937.
28. Masere, J.; Chekanov, Y.; Warren, J. R.; Stewart, F.; Al-Kaysi, R.; Rasmussen, J. K.; Pojman, J. A. *J. Polym. Sci., Part A. Polym. Chem.* **2000**, *38*, 3984.
29. Smith, S.B. U.S. Patent 6,277,939, 2001.
30. Sjardijn, W.; Snel, J.J.M. U.S. Patent 5,109,073, 1992.
31. Pojman, J.A.; Warren, J. *ACS Polym. Prepr.* **1998**, *39*(2), 356.

Chapter 11

Spontaneous Frontal Polymerization: Propagating Front Spontaneously Generated by Locally Autoaccelerated Free-Radical Polymerization

Kouichi Asakura[1], Eisuke Nihei[2], Hirokazu Harasawa[1], Akihito Ikumo[1], and Shuichi Osanai[1]

[1]Department of Applied Chemistry, Faculty of Science and Technology, Keio University, Yokohama 223–8522, Japan
[2]Department of Applied Physics and Physico-infomatics, Faculty of Science and Technology, Keio University, Yokohama 223–8522, Japan

Free-radical polymerization exhibits nonlinear phenomena, such as the Trommsdorff effect and thermal autocatalysis. A propagating front, an interface between reacted and unreacted zones, thus can be observed in the free-radical polymerization system. The frontal polymerization is usually started by attaching a mixture of monomer and initiator to an external heat source. In this chapter, we introduce that the propagating front can be generated at the center of the reaction system by the local accumulation of the heat of reaction. Polymerization of methyl methacrylate was carried out in a test tube in a thermostatic oil bath. Under optimum conditions, the front spontaneously generated at the center was propagated to complete the polymerization without forming bubbles. The phenomena might be an alternative to an interfacial gel polymerization for preparing functionally gradient materials.

There are two nonlinear phenomena in free-radical polymerizations, the Trommsdorff effect and thermal autocatalysis. In the Trommsdorff effect, the total rate of polymerization is increased as the polymerization proceeds since the diffusion-controlled termination reaction is slowed. Increasing the viscosity of the reaction system scarcely decreases the rate of the propagation reaction (*1*, *2*). Thermal autocatalysis, often called thermal runaway, also makes the polymerization strongly autoaccelerated because of the exothermic propagation reaction (*3*). Nonlinear behaviors are usually undesirable in industrial polymer production, since the polymers thus produced are inferior in quality. They tend to have large variance in molecular weight distribution and contain relatively large amounts of unreacted monomer (*4*). In addition, they sometimes result in reactor explosion.

Frontal polymerization discovered in 1972 (*5*) could be realized in free-radical polymerization because of its nonlinear behavior. If the top of a mixture of monomer and initiator in a tube is attached to an external heat source, the initiators are locally decomposed to generate radicals. The polymerization locally initiated is autoaccelerated by the combination of the Trommsdorff effect and the thermal autocatalysis exclusively at the top of the reaction system. An interface between reacted and unreacted regions, called propagating front, is thus formed. Pojman et al. extensively studied the dynamics of frontal polymerization (*6-9*) and its application in materials synthesis (*10-13*).

Koike, Nihei, and the coworkers have developed gradient refractive index (GRIN) plastic optical fiber (*14*) and plastic lens (*15*). An interfacial gel polymerization (*16*) was employed for the preparation of the GRIN polymers. They performed polymerization of methyl methacrylate (MMA) in a tube made of poly(methyl methacrylate) (PMMA). The surface of the PMMA is swollen by MMA to form a gel layer. The polymerization proceeds faster in the gel layer than in the bulk of the MMA because of the Trommsdorff effect. The isothermal front was thus formed and propagated towards the center. By adding a dopant into the polymerization system, a radial gradient in concentration of the dopant was formed and the concentration became the highest at the center, since the dopant was forced into the center by the propagating isothermal front. GRIN type polymers could be produced if the dopant changes the refractive index of the PMMA. Inserting a thin PMMA rod in the center of the polymerization system also resulted in the GRIN polymer. In this case, concentration of the dopant was the lowest at the center.

Our initial motivation to start this research project was to investigate whether a thermal front can force a dopant to make a distribution in its concentration. If the radial gradient in concentration of the dopant can be formed by thermal frontal polymerization, it might be an alternative to the interfacial gel polymerization for manufacturing the GRIN polymer. Chekanov and Pojman have succeeded in preparing PMMA having an axial gradient of dye by thermal frontal polymerization (*12*). But, this procedure requires an external pump to create the composition variation. Polymerization of MMA in a test tube immersed in a thermostatic bath was carried out at first. We thought that the propagating front would form near the wall of the test tube and propagated to the center of the reaction system. However, all attempts to generate a propagating front near the wall of the test tube failed. By choosing the appropriate bath temperature, size of test tube, and concentration of the initiator, a spherical propagating front was found to be formed at the center of the polymerization system by the local accumulation of heat of reaction. The front propagated to complete the polymerization without formation of bubbles when diphenyl sulfide was added to the reaction system. In this chapter, our research on the spontaneous frontal polymerization and its application to the preparation of transparent polymer is introduced.

Profiles of Polymerization of MMA Performed in Test Tubes

Attempts to Generate a Propagating Front Near the Wall of the Test Tube

The polymerization of MMA using 2, 2-azobis(isobutyronitrile) (AIBN) as an initiator in a test tube, having an inner diameter of 23 mm, was carried out in 60°C oil bath. The concentration of AIBN was 2.0 wt% and 15.0 mL of MMA was polymerized. In order to prevent the radical trap by the oxygen in air, argon gas flowed into the test tube at a rate of 5 mL/min. In the early stage of the reaction, heat came from the bath through the wall of the test tube to initiate the polymerization. Frontal polymerization is usually initiated at the part of the reaction system attached to the heat source (3-5). We thus expected that the front was generated near the wall of the test tube and propagated towards the center of the reaction system. If this situation can be realized, it could be regarded as the same as the interfacial gel polymerization of MMA initiated from the surface of the tube made of PMMA (16).

Contrary to our expectations, the propagating front could not be generated near the wall of the test tube. A buoyancy-driven convection prevented the formation of the front. Experiments were carried out by changing the size of the test tube and the temperature. In test tubes having inner diameters of 23, 13, or 8 mm, AIBN was dissolved in 15.0, 5.0, or 1.7 mL of MMA. The depth of MMA liquid in each test tube and the concentration of AIBN (2.0 wt%) was kept constant. The tubes were then placed in a 50°C oil bath to start the polymerization. In all cases, the front was not generated near the wall of the test tubes.

Autoacceleration of Polymerization of MMA

In order to seek alternative reaction conditions in which the propagating front could be formed, the kinetics of the polymerization were studied. Changes in conversion and temperature under each condition were monitored. Polymerization was stopped at specific times by picking up a test tube from the bath and immersing it into liquid nitrogen. The reaction mixture was then dissolved in CHCl₃ and the solution was diluted with CDCl₃ for ¹H NMR spectroscopy analysis. Conversion was determined by the ratio of integral values of proton signals of ester methyl groups of PMMA and MMA. Temperature change in the polymerization system was monitored by immersing a thermocouple into the center of each test tube. In order to characterize the PMMA produced in the reaction system, the number average molecular weight, M_n, and polydispersity, M_w/M_n, of PMMA were determined by GPC analysis.

The autoacceleration in the polymerization was observed in all cases as shown in Figure 1. Apparent autoacceleration was observed earlier stage in the reaction as the inner diameter of the test tube increased. In addition, the autoacceleration observed in the smallest test tube was not as abrupt as the one observed in the larger test tubes.

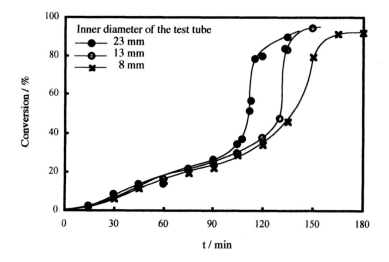

Figure 1. Conversion as a function of time for free-radical polymerization of MMA performed in test tubes having different diameters.

As shown in Figure 2, the temperature at the center of the larger test tubes suddenly rose by more than 60 degrees at the onset of the autoacceleration, while the temperature rise was only several degrees in the smallest test tube.

Both the Trommsdorff effect and thermal autocatalysis can lead to the autoacceleration in free-radical polymerization. The results indicated that the autoacceleration observed in the smallest test tube was mostly due to the Trommsdorff effect, while both the Trommsdorff effect and thermal autocatalysis strongly affected the onset of autoacceleration in the larger polymerization system. When an exothermic reaction is performed in a larger system, more heat tends to be accumulated in a reactor, since a surface-to-volume ratio is decreased as the size of the system increases. There was a critical size in the inner diameter of the test tube at which the behavior of the autoacceleration of the polymerization changes.

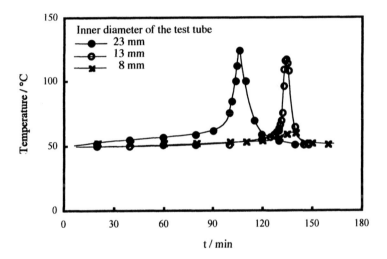

Figure 2. Temperature change at the center of a reaction system of free-radical polymerization of MMA performed in test tubes having different diameters.

The difference in the mechanism of autoacceleration, depending on the size of the test tubes, changed the character of the PMMA produced in the reaction system. As shown in Figure 3, both M_n and M_w/M_n increased in the smallest test tube after the onset of autoacceleration. This is a typical phenomenon for the Trommsdorff effect, since it is led by the retardation of the termination reaction. Increase of M_n of the polymer, however, was not observed in the larger test tubes. Thermal runaway led to rapid decomposition of the initiator to produce more radicals transiently. This process produced more polymers having a lower degree of polymerization, lowering M_n.

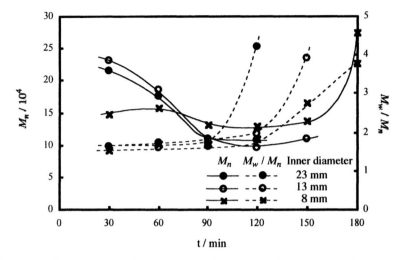

Figure 3. Changes of number average molecular weight, M_n, and polydispersity, M_w/M_n, of PMMA produced by the polymerization of MMA in test tubes having different diameters.

Thermal runaway should usually be avoided in polymer industrial plants, since it lowers the quality of the polymers and damages the reactor. It seemed that our trial to produce radial functionally gradient polymers by frontal polymerization failed.

Spontaneous Generation of a Propagating Front

Local Autoacceleration of Polymerization to Form a Propagating Front

We carefully checked the polymerization system in which the thermal runaway reaction was performed. A propagating front was found to be formed spontaneously at the center when the polymerization exhibited autoacceleration in the larger test tubes. The polymerization in a test tube having an inner diameter of 13 mm was recorded by digital video camera. No visible inhomogeneity was observed before the onset of the autoacceleration. A heated and reacted region was then generated at the center of the polymerization system to form a propagating front as shown in Figure 4.1. This indicates that thermal runaway locally occurred at the center and spread throughout the whole reaction system. Heat generated by the polymerization is diffused through the wall of the test tube. Spatial temperature distribution was thus formed to ignite the local thermal runaway at the center. The formation of the propagating front was also observed in a test tube having an inner diameter of 23 mm as shown in Figure 4.2. No visible inhomogeneity was observed when the polymerization was carried out in a test tube having an inner diameter of 8 mm.

1. Inner diameter of the test tube: 13 mm

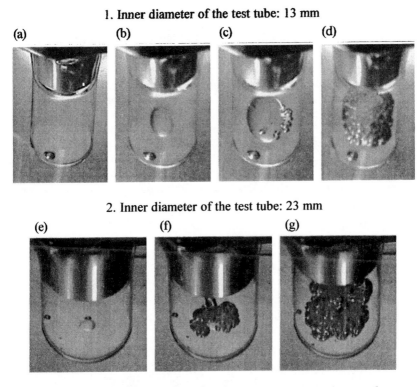

2. Inner diameter of the test tube: 23 mm

Figure 4. Spontaneous generation of the propagating front in the polymerization of MMA in the larger test tubes by setting the bath temperature at 50 °C. (a): Just before the appearance of the front; (b), (c), and (d): 10, 50, and 70 sec after the appearance of the front. (e), (f), and (g): 5, 20, 50 sec after the appearance of the front.

The shape of the propagating front could be changed by changing the depth of the MMA solution in the test tube. The longitude of the sphere increased by increasing the depth of the solution. The situation just after the spontaneous formation of the propagating front in a long tube reactor could be regarded as same as the one just after immersing a thin PMMA rod into the center of a mixture of MMA and initiator to start interfacial gel polymerization. Unlike the interfacial gel polymerization, the preparation of the tube or thin rod made of PMMA is not needed for this spontaneous frontal polymerization. This phenomenon thus has a potential importance for preparing radial functionally gradient type polymers. Bubbles, however, were formed before the completion of the reaction as shown in Figure 4. In order to prepare the transparent functionally gradient polymer by this method, the formation of the bubbles should definitely be eliminated.

Attempts to Complete Spontaneous Frontal Polymerization without Formation of Bubbles

Release of gas by the decomposition of initiator and boiling of monomer lead to the formation bubbles in free radical polymerization. Masere et al. developed gas-free initiators for high temperature polymerization (*17*). The formation of bubbles in this case, however, was mostly due to the boiling of MMA. In order to reduce the temperature near the front, polymerization in a test tube having an inner diameter of 13 mm was carried out by setting the temperature of the oil bath at 45, 50, and 55 °C. Maximum temperature during polymerization was found to be lowered by decreasing the bath temperature as shown in Figure 5.

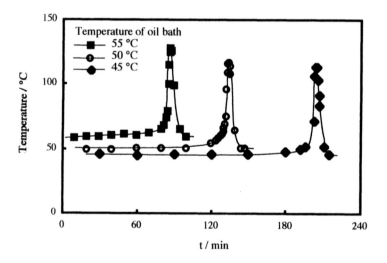

Figure 5. Temperature change at the center of the reaction system of free-radical polymerization of MMA performed in a test tube having an inner diameter of 13 mm in oil bath at different temperatures.

The formation of the bubbles, however, could not be completely inhibited by decreasing the bath temperature to 45 °C as shown in Figure 6.1. Chechilo et al. performed the frontal polymerization of MMA under high pressure to inhibit the formation of the bubbles (*18, 19*). The cost of manufacturing, however, increases if the polymerization is carried out under high pressure. In order to seek reaction conditions at which no bubbles are formed before the completion of the spontaneous frontal polymerization, the polymerization was carried out adding diphenyl sulfide. The bubble formation was completely inhibited by adding 5.0 wt% of diphenyl sulfide as shown in Figure 6.2.

1. Polymerization without adding diphenyl sulfide

(a)   (b)   (c)   (d)

2. Polymerization adding 5.0 wt% diphenyl sulfide

(e)   (f)   (g)   (h)

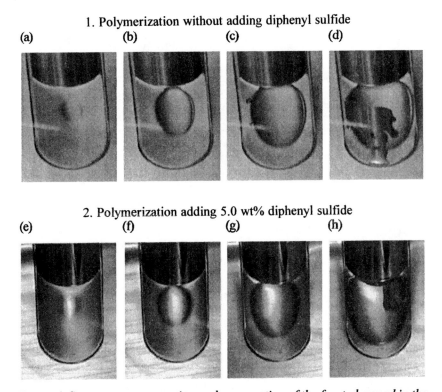

Figure 6. Spontaneous generation and propagation of the front observed in the polymerization of MMA by setting the bath temperature at 45 °C. (a), (b), (c), and (d): 10, 50, 110, and 170 sec after the appearance of the front. (e), (f), (g), and (h): 10, 40, 110, and 220 sec after the appearance of the front.

Temperature change at the center of the polymerization system containing diphenyl sulfide was monitored. As shown in Figure 7, the maximum temperature was about 10 degrees lower than the one observed in the polymerization system in the absence of diphenyl sulfide. The role of diphenyl sulfide in bubble prevention, however, has yet to be determined and the mechanism is now being in considered.

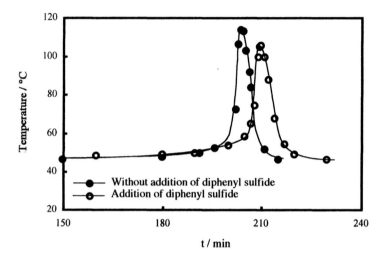

Figure 7. Temperature change at the center of the reaction system of free-radical polymerization of MMA in the presence and absence of diphenyl sulfide.

A transparent polymer rod was prepared under these conditions. The method does not require extra treatments such as preparation of a tube or thin rod made of PMMA before polymerization or reaction conditions requiring high pressure. It might be an alternative to interfacial gel polymerization for manufacturing radial functionally gradient materials.

Conclusion

The polymerizations of MMA using AIBN as an initiator in a test tube in a thermostatic oil bath were carried out by changing the size of the test tube and the temperature of the oil bath. All attempts to generate a propagating front near the wall of the test tube failed. A spherical propagating front, however, was found to be formed at the center of the reaction system when the polymerization was carried out in larger test tubes. The spontaneous polymerization was completed without formation of bubbles by setting the bath temperature at 45°C, choosing the test tube with an inner diameter of 13mm, and adding 5.0 wt% of diphenyl sulfide in the polymerization system. We are now trying to apply this spontaneous frontal polymerization for the preparation of transparent functionally gradient polymers.

Acknowledgement

This work was supported by Grant-in-Aid for the 21st Century COE program "Keio Life Conjugate Chemistry" from the Ministry of Education, Culture, Sports, Science, and Technology, Japan.

References

1. Norrish, R. G. W.; Smith, R. R. *Nature* **1942**, *150*, 336-337.
2. Trommsdorff, E.; Kohle, H.; Lagally, P. *Macromol. Chem.* **1948**, *1*, 169-198.
3. Epstein, I .R.; Pojman J. A. *An Introduction to Nonlinear Chemical Dynamics -Oscillations, Waves, Patterns, and Chaos;* Oxford Univ. Press: Oxford, 1998; pp 231-248.
4. Davtyan, S. P.; Zhirkov, P. V.; Volfson, S. A. *Russ. Chem. Rev.* **1984**, 53, 150-163.
5. Chechilo, N. M.; Khvilivitskii, R. J.; Enikolopyan, N. S. *Dokl. Akad. Nauk.* **1972**, *204*, 1180-1181.
6. Pojman, J. A.; Willis, J.; Fortenberry, D; Ilyashenko, V.; Khan, A. *J. Polym. Sci. Part A: Polym. Chem.* **1995**, *33*, 643-652.
7. Pojman, J. A.; Curtis, G.; Ilyashenko, V. M. *J. Am. Chem. Soc.* **1996**, *115*, 3783-3784.
8. Pojman, J. A.; Ilyashenko, V. M; Khan, A. M. *J. Chem. Soc. Faraday Trans.* **1996**, *92*, 2824-2836.
9. Pojman, J. A.; Elcan, W.; Khan, A. M.; Mathias, L. *J. Polym. Sci. Part A: Polym. Chem.* **1997**, *35*, 227-230.
10. Nagy, I. P.; Spike, L.; Pojman, J. A. *J. Am. Chem. Soc.* **1995**, *117*, 3611-3612.
11. Khan, A. M.; Pojman, J. A. *Trends. Polym. Sci.* **1996**, *4*, 253-257.
12. Chekanov, Y. A.; Pojman J. A. *J. Appl. Polym. Sci.* **2000**, *78*, 2398-2404.
13. Masere, J.; Lewis, L. L.; Pojman J. A. *J. Appl. Polym. Sci.* **2001**, *80*, 686-691.
14. Nihei, E.; Ishigure, T; Koike, Y. *Optimization of Modal and Material Dispersions in High-Bandwidth Graded-Index Polymer Optical Fibers, in Photonic and Optoelectronic Polymers;* Jenekhe, S. A.; Wynne, K. T., Eds.; ACS Symposium Series, American Chemical Society: Washington DC, 1997; Vol. 672, pp 58-70.

146

15. Koike, Y.; Asakawa, A, Wu, S. Nihei, E. *Apll. Opt.* **1995**, *34*, 4669-4673.
16. Koike, Y.; Takezawa, Y, Ohtsuka, Y. *Apll. Opt.* **1988**, *27*, 486-491.
17. Masere, J.; Chekanov, Y.; Warren, J. R.; Stewart, F.; Al-Kaysi, R.; Ramussen, J. K.; Pojman, J. A. *J. Polym. Sci. Part A Polym. Chem.* **2000**, *38*, 3984-3990.
18. Chechilo, N. M.; Enikolopyan, N. S. *Dokl Phys Chem* **1975**, *221*, 392-394.
19. Chechilo, N. M.; Enikolopyan, N. S. *Dokl Phys Chem* **1976**, *230*, 840-843.

Chapter 12

Bifurcation Analysis of Polymerization Fronts

D. M. G. Comissiong[1], L. K. Gross[2], and V. A. Volpert[1,*]

[1]Department of Engineering Sciences and Applied Mathematics,
Northwestern University, Evanston, IL 60208–3125
[2]Department of Theoretical and Applied Mathematics, The University
of Akron, Akron, OH 44325–4002

A free boundary model is used to describe frontal polymerization. Weakly nonlinear analysis is applied to investigate pulsating instabilities in two dimensions. The analysis produces a pair of Landau equations, which describe the evolution of the linearly unstable modes. Onset and stability of spinning and standing modes is described.

Introduction

We consider the nonlinear dynamics of a free radical polymerization front in two dimensions. Frontal polymerization (FP) refers to the process, by which conversion from monomer to polymer occurs in a narrow region that propagates in space. It was first documented experimentally by Chechilo, Khvilivitskii and Enikolopyan *(1)*. In the simplest case, a polymerization front can be generated in a test tube containing a mixture of monomer and initiator by supplying heat to one end of the tube. The heat subsequently decomposes the initiator into free radicals, which trigger the highly exothermic process of free-radical polymerization. The focus of our attention will be the self-sustaining wave which travels through the tube as polymer molecules are being formed. Uniformly propagating planar waves may become unstable as parameters vary resulting in interesting nonlinear behaviors *(2)*. To ensure the desired uniformity and quality of the resulting product, it is important to have a clear understanding of the stability of the propagating front. A complete linear stability analysis was first presented by Schult and Volpert *(3)*, and a one-dimensional nonlinear stability analysis was done later by Gross and Volpert *(4)*. Our paper extends this nonlinear analysis to two dimensions, which allows us to describe and analyze the spinning waves that have been observed experimentally and also standing waves.

Mathematical Model

Consider a cylindrical shell of circumference L, in which the reaction propagates longitudinally. In a fixed coordinate frame $(\tilde{x}, y)$ the direction of motion of the front is $-\tilde{x}$ where $-\infty < \tilde{x} < \infty$ and $0 < y < L$. By introducing a moving coordinate system $x = \tilde{x} - \varphi(y, t)$ where φ is the location of the reaction front at time t, we fix the front at $x = 0$. The dependent variables in our model are the temperature $T(x, y, t)$, monomer concentration $M(x, y, t)$, initiator concentration $I(x, y, t)$ and velocity of the propagating front $u(y, t) = -\partial\varphi(y, t)/dt$. Under sharp-interface approximation (3), (5), we solve the reactionless equations

$$\frac{\partial T}{\partial t} - \varphi_t \frac{\partial T}{\partial x} = \kappa \nabla^2 T,$$

$$\frac{\partial M}{\partial t} - \varphi_t \frac{\partial M}{\partial x} = 0,$$

$$\frac{\partial J}{\partial t} - \varphi_t \frac{\partial J}{\partial x} = 0,$$

both ahead of and behind the front. Here $J = \sqrt{I}$, κ is the thermal diffusivity and ∇^2 the Laplacian for the moving coordinate system defined as

$$\nabla^2 = \frac{\partial^2}{\partial y^2} + \left(1 + \varphi_y^2\right) \frac{\partial^2}{\partial x^2} - 2\varphi_y \frac{\partial^2}{\partial x \partial y} - \varphi_{yy} \frac{\partial}{\partial x}.$$

Boundary conditions far ahead of the front $(x \to -\infty)$ are $T \to T_0$, $M \to M_0$, $J \to J_0$, whereas far behind the front $(x \to \infty)$, $T_x \to 0$. We assume that for all $x > 0$, $J = 0$, as the initiator is completely consumed as the reaction front progresses . We assume periodic boundary conditions in y for both T and φ. The front conditions derived from sharp-interface analysis are (3), (5)

$$[T] = 0,$$

$$\kappa [T_x] = \frac{q (M_0 - M_b) \varphi_t}{1 + \varphi_y^2},$$

$$\frac{\varphi_t^2}{1 + \varphi_y^2} = F(T_b) \equiv \frac{\kappa k_{01} R_g T_b^2}{q M_0 E_1} \exp(j_0 - \frac{E_1}{R_g T_b}) \left(\int_0^{j_0} \frac{e^\eta - 1}{\eta} d\eta \right)^{-1},$$

$$M_b = f(T_b) = M_0 \exp(-j_0), \quad j_0 = \frac{J_0 A_2}{A_1}.$$

The brackets denote a jump in a quantity across the front $[v] = v(x = 0^+) - v(x = 0^-)$, q is the heat released per unit concentration of monomer, T_b and

M_b are the temperature and monomer concentration at the front respectively, $A_i = k_{0i} \exp(-E_i/R_g T_b)$ is the Arrhenius function for decomposition $(i = 1)$ and polymerization $(i = 2)$ reactions, with k_{0i} and E_i being the frequency factors and activation energies of both processes. The universal gas constant is denoted as R_g.

Basic Solution and Its Stability

The stationary solution in 1-D is

$$\widehat{T}(x) = \begin{cases} T_0 + (\widehat{T}_b - T_0)\exp(\widehat{u}x/\kappa), & x < 0 \\ \widehat{T}_b, & x > 0 \end{cases},$$

$$\widehat{M}(x) = \begin{cases} M_0, & x < 0 \\ \widehat{M}_b, & x > 0 \end{cases}, \qquad \widehat{J}(x) = \begin{cases} J_0, & x < 0 \\ 0, & x > 0 \end{cases},$$

$$\widehat{\varphi}_t = -\widehat{u} > 0,$$

where

$$\widehat{T}_b = T_0 + q(M_0 - \widehat{M}_b), \quad \widehat{M}_b = f(\widehat{T}_b), \quad \widehat{u}^2 = F(\widehat{T}_b).$$

A detailed account of the linear stability analysis of this system can be found in *(3)*, *(4)*. We use the following non-dimensional parameters in the analysis

$$z_1 = \frac{F'(\widehat{T}_b)(\widehat{T}_b - T_0)}{F(\widehat{T}_b)} \equiv 2(\widehat{T}_b - T_0)\frac{\partial \ln \widehat{u}}{\partial \widehat{T}_b}, \quad z_2 = \frac{F''(\widehat{T}_b)(\widehat{T}_b - T_0)^2}{2F(\widehat{T}_b)},$$

$$z_3 = \frac{F'''(\widehat{T}_b)(\widehat{T}_b - T_0)^3}{6F(\widehat{T}_b)}, \quad P_1 = qf'(\widehat{T}_b) \equiv q\frac{\partial \widehat{M}_b}{\partial \widehat{T}_b},$$

$$P_2 = \frac{1}{2}q(\widehat{T}_b - T_0)f''(\widehat{T}_b), \quad P_3 = \frac{1}{6}q(\widehat{T}_b - T_0)^2 f'''(\widehat{T}_b).$$

The resulting dispersion relation in non-dimensional form is *(4)*

$$4\Omega^3 + (1 + 4s^2 + 4z_1 - (z_1 - P_1)^2)\Omega^2 + z_1(1 + 4s^2 + P_1)\Omega + s^2 z_1^2 = 0.$$

Here $k = 2\pi j/L$, $j = 1, 2, \ldots$, $s = \kappa k/\widehat{u}$ is the non-dimensional wavenumber and $\Omega = \kappa \omega/\widehat{u}^2$ is the non-dimensional frequency of oscillation. Instability occurs when a pair of complex conjugate eigenvalues crosses the imaginary axis as the parameters vary. At the stability boundary

$$z_{1cr} = 4 + 2P_1 - 4s^2 \left(1 + 4s^2 + P_1\right)^{-1}$$

$$+2 \left(\left(2 + P_1 - 2s^2 \left(1 + 4s^2 + P_1\right)^{-1}\right)^2 - P_1^2 + 1 + 4s^2\right)^{\frac{1}{2}},$$

$$\omega_0^2 = (\text{Im }\Omega)^2 = \frac{1}{8} z_{1cr} \left(1 + 4s^2 + P_1\right).$$

The neutral stability curve in the (s, z_1) plane has a minimum at $s = s_m > 0$ for all $P_1 \geq 0$.

Weakly Nonlinear Analysis

We perform a nonlinear analysis that will allow us to obtain amplitude equations to characterize the evolution of the unstable modes. Gross and Volpert *(4)* have studied the 1-D case for loss of neutral stability at the wavenumber $s = 0$. This corresponds to sufficiently small values of the tube circumference L. The analysis in this instance results in a single Landau-Stuart equation which governs the weakly unstable modes. If, however, the tube circumference L is large, loss of stability will occur for some $s > 0$. Our nonlinear analysis yields a coupled set of Landau amplitude equations.

We introduce time scales $t_0 = t$, $t_1 = \epsilon t$, $t_2 = \epsilon^2 t$ and expand T, M, and φ as $T = \widehat{T} + \epsilon T_1 + \epsilon^2 T_2 + \epsilon^3 T_3 + ...$, $M = \widehat{M} + \epsilon M_1 + \epsilon^2 M_2 + \epsilon^3 M_3 + ...$, $\varphi = -\widehat{u} t_0 + \epsilon \varphi_1 + \epsilon^2 \varphi_2 + \epsilon^3 \varphi_3 + ...$ Also $z_1 = z_{1cr} + \mu \epsilon^2$. To non-dimensionalize the above the following scales are used , where $j = 0, 1, 2$

$$\xi = \frac{\widehat{u}}{\kappa} x, \quad \eta = \frac{\widehat{u}}{\kappa} y, \quad \psi = \frac{\widehat{u}}{\kappa} \varphi, \quad \psi_j = \frac{\widehat{u}}{\kappa} \varphi_j, \quad t_j = \frac{\kappa}{\widehat{u^2}} \tau_j, \quad \theta = \frac{T}{\widehat{T_b} - T_0},$$

$$\theta_b = \frac{T_b}{\widehat{T_b} - T_0}, \quad m = \frac{M}{-\widehat{M_b} + M_0}, \quad m_b = \frac{M_b}{-\widehat{M_b} + M_0}, \quad m_{jb} = \frac{M_{jb}}{-\widehat{M_b} + M_0}.$$

Consequently, we have the following sequence of problems

$$\frac{\partial \theta_j}{\partial \tau_0} + \frac{\partial \theta_j}{\partial \xi} - \left(\frac{\partial^2 \theta_j}{\partial \xi^2} + \frac{\partial^2 \theta_j}{\partial \eta^2}\right) - \left(\frac{\partial \psi_j}{\partial \tau_0} + \frac{\partial^2 \psi_j}{\partial \eta^2}\right) \frac{d\widehat{\theta}}{d\xi} = \widetilde{Q}_j, \tag{1}$$

$$[\theta_j] = 0, \quad \left[\frac{\partial \theta_j}{\partial \xi}\right] - \frac{\partial \psi_j}{\partial \tau_0} - m_{jb} = \widetilde{R}_j, \tag{2}$$

$$2\frac{\partial \psi_j}{\partial \tau_0} + z_{1cr}\theta_{jb} = \widetilde{S}_j, \quad m_{jb} - P_1\theta_{jb} = \widetilde{T}_j, \tag{3}$$

where $\widetilde{Q}_j$, $\widetilde{R}_j$, $\widetilde{S}_j$ and $\widetilde{T}_j$ are given in the Appendix. The solution θ_j, ψ_j satisfy periodic boundary conditions in η and

$$\frac{\partial \theta_j}{\partial \xi}\bigg|_{\xi=+\infty} = 0, \quad \theta_j|_{\xi=-\infty} = 0, \tag{4}$$

where $\widetilde{Q}_j$, $\widetilde{R}_j$, $\widetilde{S}_j$ and $\widetilde{T}_j$ are given in the Appendix.
The solvability condition for problem (1)-(4) is

$$\int_0^{\frac{2\pi}{\omega_0}} \int_0^L \left(\bar{v}|_{\xi=0} \left(\frac{P_1}{z_{1cr}}\widetilde{S}_j + \widetilde{R}_j + \widetilde{T}_j \right) - \frac{1}{z_{1cr}}\widetilde{S}_j \left[\frac{\partial \bar{v}}{\partial \xi} \right] \right) d\eta d\tau_0$$

$$= \int_0^{\frac{2\pi}{\omega_0}} \int_0^L \int_{-\infty}^{\infty} \widetilde{Q}_j \bar{v} d\tau_0 d\eta d\xi,$$

where $\bar{v}$ is a solution of the adjoint problem

$$v_{\pm}(\xi, \eta, \tau_0) = \begin{cases} \exp\left(i\omega_0\tau_0 \pm is\eta + \frac{1}{2}\left(-1 + \bar{d}\right)\xi \right), & \xi < 0 \\ \exp\left(i\omega_0\tau_0 \pm is\eta + \frac{1}{2}\left(-1 - \bar{d}\right)\xi \right), & \xi > 0 \end{cases},$$

$$v_0(\xi) = \begin{cases} 1, & \xi < 0 \\ \exp(-\xi), & \xi > 0 \end{cases}.$$

The $O(\epsilon)$ Problem (j=1)

The solution of the $O(\epsilon)$ problem is

$$\theta_1 = \left(Ae^{i(\omega_0\tau_0+s\eta)} + Be^{i(\omega_0\tau_0-s\eta)} \right) X_1(\xi) + CC,$$

$$\psi_1 = \frac{Az_{1cr}}{2}e^{i(\omega_0\tau_0+s\eta)} + \frac{Bz_{1cr}}{2}e^{i(\omega_0\tau_0-s\eta)} + CC + \mho,$$

where CC denotes the complex conjugate, $\mho$, A and B are functions of the slow times and

$$X_1(\xi) = \begin{cases} -\left(i\omega_0 + \frac{1}{2}z_{1cr}\right)e^{\frac{1}{2}(1+d)\xi} + \frac{1}{2}z_{1cr}e^{\xi}, & \xi < 0 \\ -i\omega_0 e^{\frac{1}{2}(1-d)\xi}, & \xi > 0 \end{cases}.$$

The $O\left(\epsilon^2\right)$ Problem (j=2)

Applying the solvability condition to the $O\left(\epsilon^2\right)$ problem with $v = v_+$ and $v = v_-$ shows that A and B depend only on the slow time τ_2. When $v = v_0$ the solvability condition yields $\partial \mho/d\tau_1 = r_0\left(|A|^2 + |B|^2\right)$ with r_0 as

$$
r_0 = \frac{z_{1cr}}{-2(P_1 + 1)}\left\{-2\omega_0^2\left(\frac{P_1 + 1}{z_{1cr}}\left(\frac{1}{4}z_{1cr}^2 - z_2\right) + \frac{1}{2}z_{1cr}P_1 + P_2\right)\right.
$$

$$
-\frac{s^2 z_{1cr}^2}{2}\left(1 - \frac{P_1 + 1}{z_{1cr}}\right) - \frac{z_{1cr}\omega_0^2}{2}\left(2 + \frac{1 - \overline{d}}{1 + \overline{d}} + \frac{1 - d}{1 + d}\right)
$$

$$
\left.-\frac{s^2 z_{1cr}i\omega_0}{2}\left(\frac{1 - \overline{d}}{1 + \overline{d}} - \frac{1 - d}{1 + d}\right) + \frac{s^2}{2}z_{1cr}^2\right\}.
$$

The solution for the $O\left(\epsilon^2\right)$ problem is

$$
\theta_2 = g_0\left(\xi\right)\left(|A|^2 + |B|^2\right) + \{g_1\left(\xi\right)\left(A^2\exp(2i\left(\omega_0\tau_0 + s\eta\right))\right)
$$
$$
+ B^2\exp\left(2i\left(\omega_0\tau_0 - s\eta\right)\right))
$$
$$
+ g_2\left(\xi\right)AB\exp\left(2i\omega_0\tau_0\right) + g_3\left(\xi\right)A\overline{B}\exp\left(2is\eta\right) + CC\},
$$

$$
\psi_2 = C_0\left(|A|^2 + |B|^2\right) + \{C_1(A^2\exp\left(2i\left(\omega_0\tau_0 + s\eta\right)\right)
$$
$$
+ B^2\exp\left(2i\left(\omega_0\tau_0 - s\eta\right)\right))
$$
$$
+ C_2 AB\exp\left(2i\omega_0\tau_0\right) + C_3 A\overline{B}\exp\left(2is\eta\right) + CC\},
$$

where the functions $g_j\left(\xi\right)$ are

$$
g_0 = \begin{cases} D_{01}\exp\left(\xi\right) + D_{02}\left(\exp\left(\xi\right) + \xi\exp\left(\xi\right)\right) \\ \quad +(D_{03}\exp((1 + d)\frac{\xi}{2}) + CC), \ \xi < 0 \\ a_{02} + (D_{04}\exp((1 - d)\frac{\xi}{2}) + CC), \ \xi > 0 \end{cases}
$$

$$
g_1 = \begin{cases} a_{11}\exp((1 + d_1)\frac{\xi}{2}) + D_{11}\exp(\xi) + D_{13}\exp((1 + d)\frac{\xi}{2}), \ \xi < 0, \\ a_{12}\exp((1 - d_1)\frac{\xi}{2}) + D_{14}\exp((1 - d)\frac{\xi}{2}), \ \xi > 0 \end{cases}
$$

$$g_2 = \begin{cases} a_{21} \exp((1+d_2)\frac{\xi}{2}) + D_{21}\exp(\xi) + D_{23}\exp((1+d)\frac{\xi}{2}), & \xi < 0, \\ a_{22}\exp((1-d_2)\frac{\xi}{2}) + D_{24}\exp((1-d)\frac{\xi}{2}), & \xi > 0 \end{cases}$$

$$g_3 = \begin{cases} a_{31}\exp((1+d_3)\frac{\xi}{2}) + D_{31}\exp(\xi) \\ \quad + (D_{33}\exp((1+d)\frac{\xi}{2}) + CC), & \xi < 0, \\ a_{32}\exp((1-d_3)\frac{\xi}{2}) + (D_{34}\exp((1-d)\frac{\xi}{2}) + CC), & \xi > 0 \end{cases}$$

Here, $d_1 = \sqrt{1 + 8i\omega_0 + 16s^2}$, $d_2 = \sqrt{1 + 8i\omega_0}$ and $d_3 = \sqrt{1 + 16s^2}$ and the coefficients a_{ij}, D_{ij} and C_j are given in the Appendix.

The $O\left(\epsilon^3\right)$ Problem (j=3)

The solvability conditions for the $O\left(\epsilon^3\right)$ problem yield a coupled set of Landau equations

$$\frac{\partial A}{\partial \tau_2} = \mu A\chi + A^2\overline{A}\beta_1 + AB\overline{B}\beta_2, \quad \frac{\partial B}{\partial \tau_2} = \mu B\chi + B^2\overline{B}\beta_1 + AB\overline{A}\beta_2. \quad (5)$$

The complex coefficients β_1, β_2 and χ are given in the Appendix.

Analysis of the Amplitude Equations

We let the amplitudes A and B of ψ_1, which determines the shape of the front, be of the form

$$A\left(\tau_2\right) = a\left(\tau_2\right)\exp\left(i\theta_a\tau_2\right), \quad B\left(\tau_2\right) = b\left(\tau_2\right)\exp\left(i\theta_b\tau_2\right). \quad (6)$$

Substituting (6) into (5) and separating real and imaginary parts results in

$$\frac{da}{d\tau_2} = \mu\chi_r a + \beta_{1r}a^3 + \beta_{2r}ab^2, \quad \frac{d\theta_a}{d\tau_2} = \mu\chi_i + \beta_{1i}a^2 + \beta_{2i}b^2, \quad (7)$$

$$\frac{db}{d\tau_2} = \mu\chi_r b + \beta_{1r}b^3 + \beta_{2r}a^2 b, \quad \frac{d\theta_b}{d\tau_2} = \mu\chi_i + \beta_{1i}b^2 + \beta_{2i}a^2. \quad (8)$$

Here the subscripts r and i represent the real and imaginary parts of the respective coefficients. In order to determine the steady state solutions of (7), (8) which, in the original problem, correspond to a superposition of waves traveling along the front, we consider $da/d\tau_2 = db/d\tau_2 = 0$. This leads to

$$a\left(\mu\chi_r + \beta_{1r}a^2 + \beta_{2r}b^2\right) = 0, \quad b\left(\mu\chi_r + \beta_{1r}b^2 + \beta_{2r}a^2\right) = 0. \quad (9)$$

154

There are four critical points

$$a_1 = b_1 = 0, \quad a_2 = 0, \ b_2 = w_t, \quad a_3 = w_t, \ b_3 = 0, \quad a_4 = b_4 = w_s,$$

where

$$w_t = (-\mu\chi_r/\beta_{1r})^{1/2}, \quad w_s = (-\mu\chi_r/(\beta_{1r} + \beta_{2r}))^{1/2}.$$

In the case of the first critical point the amplitudes A and B are identically equal to zero, which corresponds to the uniformly propagating wave in the original problem. The second and third critical points correspond to waves traveling along the front, which are right- and left-traveling waves, respectively. The last critical point corresponds to a standing wave.

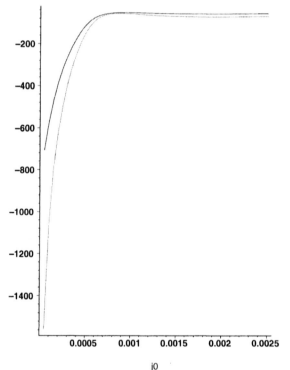

Figure 1. Graphs of β_{1r} (upper curve) and β_{2r} (lower curve) versus the non-dimensional $j_0 = J_0 \frac{A_2}{A_1}$

It can be shown that for all parameter values χ_r is positive. Thus, from the expression for w_t we conclude that left- and right-traveling wave exist for $\mu > 0$ (the so-called supercritical bifurcation) if $\beta_{1r} < 0$ and for $\mu < 0$ (the subcritical bifurcation) if $\beta_{1r} > 0$. In a similar way, the supercritical bifurcation of standing waves occurs if $\beta_{1r} + \beta_{2r} < 0$ and subcritical bifurcation occurs if $\beta_{1r} + \beta_{2r} > 0$. All the subcritical bifurcations are known to produce locally unstable regimes. The supercritical bifurcation can lead to either stable or unstable solutions depending on the parameter values. Specifically, the supercritical bifurcation of traveling waves (which occurs if $\beta_{1r} < 0$) is stable if $\beta_{2r} < \beta_{1r}$ and unstable otherwise. The supercritical bifurcation of standing waves (which occurs if $\beta_{1r} + \beta_{2r} < 0$) is stable if $\beta_{2r} > \beta_{1r}$ and unstable otherwise.

The quantities β_{1r} and β_{2r} are plotted in Fig. 1 as functions of j_0 (which is proportional to the initial concentration of the initiator) for typical parameter values (3) $(E_1 - E_2)/(R_g q M_0) = 19.79$ and $E_1/(R_g q M_0) = 58.4$. We set the wavenumber $s = 0.55$, which is close to the value s_m at which the neutral stability curve has a minimum (4). We see that both quantities are negative, which implies that both traveling and standing waves appear as a result of a supercritical bifurcation, and that for the parameter values chosen $\beta_{1r} > \beta_{2r}$, so that the traveling waves are stable while the standing waves are unstable. This observation agrees with the experimental data in (2) where spinning waves have been observed.

Acknowledgements

D.M.G. Comissiong has been supported by a Fulbright fellowship administered by LASPAU. L.K. Gross acknowledges support from the National Science Foundation (Grant DMS-0074965), the Ohio Board of Regents (Research Challenge Grant), and the Buchtel College of Arts and Sciences at the University of Akron (matching grant). V.A. Volpert has been supported in part by the National Science Foundation (Grant DMS-0103856).

Appendix

The right-hand side of equations (1)-(3) are

$$\widetilde{Q}_1 = 0, \quad \widetilde{Q}_2 = \frac{\partial}{\partial \tau_1}\left(\psi_1 \frac{d\widehat{\theta}}{d\xi} - \theta_1\right) + \left(\frac{\partial \psi_1}{\partial \tau_0} - \frac{\partial^2 \psi_1}{\partial \eta^2}\right)\frac{\partial \theta_1}{\partial \xi}$$

$$+ \frac{\partial \psi_1}{\partial \eta}\frac{\partial}{\partial \xi}\left(\frac{\partial \psi_1}{\partial \eta}\frac{d\widehat{\theta}}{d\xi} - 2\frac{\partial \theta_1}{\partial \eta}\right)$$

$$\widetilde{Q}_3 = \left(\frac{\partial\psi_2}{\partial\tau_0} - \frac{\partial^2\psi_2}{\partial\eta^2}\right)\frac{\partial\theta_1}{\partial\xi} + \frac{\partial}{\partial\tau_1}\left(\psi_2\frac{d\widehat{\theta}}{d\xi} - \theta_2\right) + \frac{\partial}{\partial\tau_2}\left(\psi_1\frac{d\widehat{\theta}}{d\xi} - \theta_1\right)$$

$$+ \frac{\partial\psi_1}{\partial\eta}\frac{\partial}{\partial\xi}\left(\frac{\partial\psi_1}{\partial\eta}\frac{\partial\theta_1}{\partial\xi} - 2\frac{\partial\theta_2}{\partial\eta}\right) + \frac{\partial\psi_1}{\partial\tau_1}\frac{\partial\theta_1}{\partial\xi}$$

$$+ 2\frac{\partial\psi_2}{\partial\eta}\frac{\partial}{\partial\eta}\left(\psi_1\frac{d^2\widehat{\theta}}{d\xi^2} - \frac{\partial\theta_1}{\partial\xi}\right) + \left(\frac{\partial\psi_1}{\partial\tau_0} - \frac{\partial^2\psi_1}{\partial\eta^2}\right)\frac{\partial\theta_2}{\partial\xi},$$

$$\widetilde{R}_1 = 0, \quad \widetilde{R}_2 = \left(\frac{\partial\psi_1}{\partial\eta}\right)^2 + \frac{\partial\psi_1}{\partial\tau_1} - m_{1b}\frac{\partial\psi_1}{\partial\tau_0},$$

$$\widetilde{R}_3 = 2\frac{\partial\psi_1}{\partial\eta}\frac{\partial\psi_2}{\partial\eta} - \frac{\partial\psi_1}{\partial\tau_0}\left(\frac{\partial\psi_1}{\partial\eta}\right)^2 + \frac{\partial\psi_2}{\partial\tau_1} + \frac{\partial\psi_1}{\partial\tau_2}$$

$$- m_{1b}\left(\left(\frac{\partial\psi_1}{\partial\eta}\right)^2 + \frac{\partial\psi_2}{\partial\tau_0} + \frac{\partial\psi_1}{\partial\tau_1}\right) - m_{2b}\frac{\partial\psi_1}{\partial\tau_0},$$

$$\widetilde{S}_1 = 0, \quad \widetilde{S}_2 = -2\frac{\partial\psi_1}{\partial\tau_1} - z_2\left(\theta_{1b}\right)^2 - \left(\frac{\partial\psi_1}{\partial\eta}\right)^2 + \left(\frac{\partial\psi_1}{\partial\tau_0}\right)^2,$$

$$\widetilde{S}_3 = -2\frac{\partial\psi_2}{\partial\tau_1} - 2\frac{\partial\psi_1}{\partial\tau_2} + 2\frac{\partial\psi_1}{\partial\tau_0}\left(\frac{\partial\psi_2}{\partial\tau_0} + \frac{\partial\psi_1}{\partial\tau_1}\right) - 2z_2\theta_{1b}\theta_{2b}$$

$$- z_3\left(\theta_{1b}\right)^3 - \mu\theta_{1b} - z_{1cr}\theta_{1b}\left(\frac{\partial\psi_1}{\partial\eta}\right)^2 - 2\frac{\partial\psi_1}{\partial\eta}\frac{\partial\psi_2}{\partial\eta},$$

$$\widetilde{T}_1 = 0, \quad \widetilde{T}_2 = P_2\left(\theta_{1b}\right)^2, \quad \widetilde{T}_3 = 2P_2\theta_{1b}\theta_{2b} + P_3\left(\theta_{1b}\right)^3.$$

The coefficients a_{ij}, D_{ij} and C_i appearing in the expressions for $g_i\left(\xi\right)$ are

$$D_{02} = -r_0, \quad D_{03} = \left(s^2 + i\omega\right)z_{1cr}(i\omega_0 + z_{1cr}/2)(1 - d)^{-1},$$

$$D_{04} = z_{1cr}\left(s^2 + i\omega_0\right)i\omega_0(1 + d)^{-1},$$

$$D_{01} = -(1 + P_1)^{-1}\{2\omega_0^2 P_2 + s^2 z_{1cr}^2/2 + P_1\omega^2 z_{1cr} - ((1 - d)D_{04}/2 + CC) + ((1 + d)D_{03}/2 + CC) + P_1\left(D_{03} + CC\right)\},$$

$$a_{02} = D_{01} + D_{03} + \overline{D_{03}} - D_{04} - \overline{D_{04}}.$$

$$D_{13} = -(i\omega_0 + z_{1cr}/2)\,(1+d)\,z_{1cr}/4, \; D_{14} = -i\omega_0 z_{1cr}\,(1-d)\,/4,$$

$$
\begin{aligned}
a_{12} = &\{D_{14}((d+d_1)\,/2 + P_1) + D_{13}\,(d-d_1)\,/2 + z_{1cr}^2(1-d_1)/16 \\
&- \omega_0^2 P_2 - s^2 z_{1cr}^2/4 - P_1\omega_0^2 z_{1cr}/2 - (\omega_0^2 z_2 + s^2 z_{1cr}^2/4 \\
&- z_{1cr}^2\omega_0^2/4 - z_{1cr}D_{14})(4i\omega_0)^{-1}(0.5 - 0.5d_1 + 2i\omega_0)\}\{(1-d_1)/2 \\
&- (1 + z_{1cr}/4i\omega_0)(1+d_1)/2 + z_{1cr}/4i\omega_0 + z_{1cr}/2 - P_1\}^{-1}
\end{aligned}
$$

$$C_1 = \{\omega_0^2 z_2 + s^2 z_{1cr}^2/4 - z_{1cr}^2\omega_0^2/4 - z_{1cr}a_{12} - z_{1cr}D_{14}\}(4i\omega_0)^{-1},$$

$$D_{11} = C_1 + z_{1cr}^2/8\,, \; a_{11} = a_{12} + D_{14} - D_{11} - D_{13}.$$

$$D_{23} = -0.5z_{1cr}(1+d)(i\omega_0 + 0.5z_{1cr}) = 2D_{13},$$

$$D_{24} = -0.5z_{1cr}i\omega_0(1-d) = 2D_{14},$$

$$
\begin{aligned}
a_{22} = &\{0.5(d - d_2)D_{23} + D_{24}\{0.5(d+d_2) + P_1 - z_1(4i\omega_0)^{-1}(1 + 2i\omega_0 \\
&- 0.5(1+d_2))\} - z_{1cr}^2(d_2 - 1)/8 + 0.5s^2 z_{1cr}^2 - P_1 z_1\omega_0^2 - 2P_2\omega_0^2 \\
&+ (1 + 2i\omega_0 - (1+d_2)/2)(2\omega_0^2(z_2 - 0.25z_{1cr}^2) - 0.5s^2 z_{1cr}^2)(4i\omega_0)^{-1}\}/ \\
&\{-d_2 + z_{1cr}(1 + 2i\omega_0 - 0.5(1+d_2))/4i\omega_0 - P_1\}
\end{aligned}
$$

$$C_2 = (-z_{1cr}a_{22} - z_{1cr}D_{24} - s^2 z_{1cr}^2/2 + 2\omega_0^2 z_2 - z_{1cr}^2\omega_0^2/2)/4i\omega_0,$$

$$D_{21} = C_2 + z_{1cr}^2/4, \; D_{33} = D_{13}, \; D_{34} = D_{14},$$

$$
\begin{aligned}
a_{21} = &\,a_{22}\,(1 + z_{1cr}/4i\omega_0) + D_{24} - z_{1cr}^2/4 - D_{23} \\
&- (-z_{1cr}D_{24} - s^2 z_{1cr}^2/2 + 2\omega_0^2 z_2 - z_{1cr}^2\omega_0^2/2)/4i\omega_0.
\end{aligned}
$$

$$a_{32} = s^2 z_{1cr}/2 - 2\omega_0^2 z_2/z_{1cr} + z_{1cr}\omega_0^2/2 - D_{34} - \overline{D_{34}},$$

$$
\begin{aligned}
a_{31} = &\{-a_{32}\,((d_3 + 1)\,/2 + P_1) - (D_{34}\,((d+1)\,/2 - P_1) + CC) \\
&+ (D_{33}\,(1-d)\,/2 + CC) + s^2 z_{1cr}^2/2 - 2\omega_0^2 P_2 - P_1\omega_0^2 z_{1cr}\}\,(d_3 - 1)\,/2,
\end{aligned}
$$

$$C_3 = a_{32} - a_{31} - z_{1cr}^2/4 - D_{33} - \overline{D_{33}} + D_{34} + \overline{D_{34}}, \; D_{31} = C_3 + z_{1cr}^2/4.$$

The coefficients in the amplitude equations (5)) are

$$\chi = i\omega_0\,(P_1 + d)\,\{z_{1cr}(P_1 + d - z_{1cr}/2 + 2i\omega_0/d + z_{1cr}/(2d))\}^{-1},$$

$$\beta_1 = (Q_5 - Q_1 - Q_3)\,\{P_1 + d - z_{1cr}/2 + 2i\omega_0/d + z_{1cr}/(2d)\}^{-1},$$

$$\beta_2 = (Q_6 - Q_2 - Q_4)\,\{P_1 + d - z_{1cr}/2 + 2i\omega_0/d + z_{1cr}/2d\}^{-1},$$

$$Q_1 + Q_3 = 2i\omega_0 C_1 I_3 + r_0 I_2 - 0.25z_{1cr}^2 s^2 I_5 + 0.5z_{1cr}^2 s^2 I_4 - 2z_{1cr} s^2 I_7$$
$$+ 2z_{1cr} s^2 C_1 I_1 + 0.5z_{1cr}(i\omega_0 + s^2)I_6 + 0.5z_{1cr}(-i\omega_0 + s^2)I_7,$$

$$Q_2 + Q_4 = 2i\omega_0 C_2 I_3 + r_0 I_2 + 0.5z_{1cr}^2 s^2 I_5 - 2z_{1cr} s^2 I_9 + 2z_{1cr} s^2 C_3 I_1$$
$$+ 0.5z_{1cr}(i\omega_0 + s^2)(I_6 + I_9) + 0.5z_{1cr}(-i\omega_0 + s^2)I_7,$$

$$I_1 = 2/(1 + d),$$

$$I_2 = -i\omega_0/d + 0.25z_{1cr}(d - 1)\{1/d - 2/(1 + d)\},$$

$$I_3 = 2i\omega_0/(d + \overline{d}) + 0.5z_{1cr}(d - 1)\{1/(d + \overline{d}) - 1/(1 + d)\},$$

$$I_4 = -0.5i\omega_0(d + 1/d) + 1/8z_{1cr}(d - 1)^3/\{d(d + 1)\} - 0.25z_{1cr}(d - 1),$$

$$I_5 = i\omega_0(\overline{d} - d) + 0.25z_{1cr}(1 - \overline{d}) + i\omega_0(d^2 + 1)/(d + \overline{d})$$
$$+ 0.25z_{1cr}(d - 1)^2\{1/(d + 1) - 1/(d + \overline{d})\},$$

$$I_6 = -0.5(d - 1)\{2D_{01}/(d + 1) - 4D_{02}/(d + 1)^2 + D_{03}/d + 2\overline{D_{03}}/(d + \overline{d})\}$$
$$+ 0.5(d + 1)\{2a_{02}/(d + 1) + D_{04}/d + 2\overline{D_{04}}/(d + \overline{d})\},$$

$$I_7 = -0.5(d - 1)\{2a_{11}/(d + d_1) + 2D_{11}/(d + 1) + D_{13}/d\}$$
$$+ 0.5(d + 1)\{2a_{12}/(d + d_1) + D_{14}/d\},$$

$$I_8 = -0.5(d-1)\{2a_{21}/(d+d_1) + 2D_{21}/(d+1) + D_{23}/d\}$$
$$+ 0.5(d+1)\{2a_{22}/(d+d_1) + D_{24}/d\},$$

$$I_9 = -0.5(d-1)\{2a_{31}/(d+d_3) + 2D_{31}/(d+1) + D_{33}/d + 2\overline{D_{33}}/(d+\bar{d})\}$$
$$+ 0.5(d+1)\{2a_{32}/(d+d_3) + D_{34}/d + 2\overline{D_{34}}/(d+\bar{d})\},$$

$$Q_5 = (P_1 + d)/z_{1cr}\{i\omega_0 z_{1cr} r_0 + 2iz_2\omega_0 g_0(0) + 2\omega_0^2 z_{1cr} C_1$$
$$+ 3i\omega_0 z_{1cr}^3 s^2/4 - 2s^2 z_{1cr} C_1 + 3iz_3\omega_0^3 - 2iz_2\omega_0 g_1(0)\}$$
$$- i\{P_1\omega_0(-8r_0 + 4z_{1cr}g_0(0) - 4z_{1cr}g_1(0)$$
$$- 6s^2 z_{1cr}^2) + 16iC_1\left(P_1\omega_0^2 + s^2 z_{1cr}\right) + 3\omega_0 z_{1cr}\left(s^2 z_{1cr}^2 + 4\omega_0^2 P_2\right)\}/8$$
$$- 2iP_2\omega_0 g_0(0) + 2iP_2\omega_0 g_1(0) - 3iP_3\omega_0^3,$$

$$Q_6 = (P_1 + d)/z_{1cr}\{2i\omega_0 z_2\left(g_3(0) - g_2(0) + g_0(0)\right) + 6i\omega_0^3 z_3$$
$$+ 2\omega_0^2 z_{1cr} C_2 + i\omega_0 z_{1cr} r_0 - 2s^2 z_{1cr} C_3 - i\omega_0 z_{1cr}^3 s^2/2\}$$
$$- i\{4\omega_0 z_{1cr} P_1\left(g_0(0) - g_2(0) + g_3(0) + s^2 z_{1cr}\right) + 16is^2 z_{1cr} C_3$$
$$- 24\omega_0^3 z_{1cr} P_2 - 2\omega_0 z_{1cr}^3 s^2 - 8P_1\omega_0 r_0 + 16iP_1\omega_0^2 C_2\}/8$$
$$+ 2i\omega_0 P_2\left(g_2(0) - g_0(0) - g_3(0)\right) - 6iP_3\omega_0^3.$$

References

1. Chechilo, N. M.; Khvilivitskii, R. J.; Enikolopyan, N. S. *Dokl. Akad. Nauk USSR* **1972**, *204*, 1180-1181.

2. Pojman, J. A.; Ilyashenko, V. M.; Khan, A. M. *Physica D* **1995**, *84*, 260-268.

3. Schult, D. A.; Volpert, V. A. *Int. J. of SHS* **1999**, *8*, 417-440.

4. Gross, L. K. ; Volpert, V. A. *Stud. Appl. Math*, in press

5. Spade, C. A.; Volpert, V. A. *Combust. Theory Modelling* **2001**, *5*, 21-39.

Chapter 13

"Cold" Ignition of Combustionlike Waves of Cryopolymerization and Other Reactions in Solids

Alain Pumir[1,*], Viktor V. Barelko[2], Igor M. Barkalov[2], and Dmitriy P. Kiryukhin[2]

[1]Institut Non Lineaire de Nice 1361, Route des Lucioles, 06560, Valbonne, France
[2]Institute of Problems of Chemical Physics, RAS, Chernogolovka, Russia

Fronts of chemical weakly exothermal reactions (such as polymerization,chorination of hydrocarbons, and others) may propagate at very low temperatures (4K - 77 K) in solids, due to a very unusual mechanism, involving a feedback between exothermal chemical reaction and disruption of the solid matrix. The reaction may then be induced by mechanical constraints, without any elevation of temperature. On the basis of a simple phenomenological model, we investigate "cold" ignition of a propagating front by initially disrupting a localized zone of the solid matrix. The main result of this work, confirmed exerimentally, is that reaction can be initiated by disrupting only a very small fraction of the sample.

Introduction

For different classes of reactions, such as polymerisation, copolymerisation, hydrocarbon halogenation, and others, propagation of waves of chemical transformations may be observed, even at very low temperatures (4K-77K) (*1*).

The propagation mechanism at such temperatures cannot result from the usual mechanism observed in standard combustion. Indeed, in classical combustion theory, the heat released by the reaction induces a temperature elevation ahead of the reaction zone, thus increasing the Arrhenius factor : $e^{-E/RT}$, where E is the energy of activation, and inducing reaction further ahead. For weakly exothermal reactions, and at very low temperatures, the Arrhenius factor remains always very small, so reactions should have an extremely low rate, very close to zero, contrary to what is observed experimentally. It was shown instead that the mechanism of propagation actually depends on the rupture of the solid matrix ahead of the reaction zone, induced by the release of thermal energy. Indeed, the heat released by the reaction leads to the formation of large temperature gradients which rupture and disperse the reagent. The reaction may proceed in this ruptured zone, thus permitting the propagation of a wave of chemical reaction.

In this process, temperature plays a very different role, compared to usual combustion theory. In particular, it was observed experimentally that the wave of chemical transformation can be initiated by applying a mechanical constraint to the sample. The purpose of the present contribution is to provide a theoretical analysis of the process of ignition of chemical waves by a localized mechanical perturbation. In particular, we estimate the critical size of the zone one needs to perturb in order to start a chemical waves.The theoretical prediction was confirmed by an experimental study, where "cold" ignition of travelling wave was realized by applying a local perturbation, over a limited region of space, in practice, by pricking or scratching the solid matrix with a thin needle. Applications of the concept to mechanisms of chemical cosmology and to problem of initiation of solid explosives by friction or shock is briefly discussed.

Theoretical Analysis

As explained in the introduction, it is assumed that reaction can proceed only when the solid matrix becomes dispersed. In the model proposed in

(*1*), brittle fracture is postulated to result from a thermal shock : a temperature gradient larger than a critical value induces a disruption of the matrix, and chemical reaction may start, therefore leading to heat release for a time τ. The model reads :

$$\partial_t T = D\partial_x^2 T + Q \tag{1}$$

$Q = 0$ as long as $\left|\partial_x T\right| < \left(\partial_x T\right)_c$

$Q = Q_0$ for a time τ after $\left|\partial_x T\right|$ has reached $\left(\partial_x T\right)_c$ $\tag{2}$

In this model, one may determine travelling wave solutions, by solving :

$$-v\partial_\xi T = D\partial_\xi^2 T + Q \tag{3}$$

with $\xi = x - vt$. The solution can be easily determined. It is piecewise exponential. By imposing the proper boundary conditions at $\pm\infty$ one obtains (*1*) :

$$G = (1 - \exp(-U^2))\,/\,U \tag{4}$$

where :

$$U = v(\tau/D)^{1/2} \quad \text{and} \quad G = ((\partial_x T)_c\,/\,Q_0)(D\,/\,\tau)^{1/2} \tag{5}$$

The parameter G provides a dimensionless measure of the critical thermal stress, divided by the amount of released heat during the reaction. Two solutions exist when $G \leq G_c \approx 0.64$. In the limit $G \to 0$, one solution has a low velocity : $v \approx D(\partial_x T)_c\,/\,(Q_0\tau)$ and another one has a fast velocity : $v \approx Q_0\,/\,(\partial_x T)_c$.

To understand the generation of waves when a localized fractured region is introduced at t = 0, we notice that the reaction will generate heat ($Q\neq0$ for a time τ after the material has been fractured). As a result of this localized (inhomogeneous) heat release, a flux is created, hence the temperature derivative becomes non zero. The heat released, due to the disruption of the matrix at time t = 0 is proportional to the product of Q_0 by the size of the fractured region, denoted by Δ. When this heat release is large enough, then the temperature gradient may reach the critical value, and a chemical wave may start. The temperature elevation resulting from this localized release of energy can be computed from Eq.(1) :

$$T(x,t) = \int_0^t \frac{dt'}{\sqrt{4\pi(t-t')}} \int_{-\Delta/2}^{+\Delta/2} Q(x',t') \exp\left(-\frac{(x-x')^2}{4D(t-t')}\right) dx' \quad (6)$$

The localized character of the source term is explicitly taken into account in the equation above. The fact that the highest value of the heat flux is reached at the boundary of the fractured zone is physically reasonable, and can be justified by a precise analysis. As long as the threshold value $(\partial_x T)_c$ has not been reached before, the temperature derivative at x = Δ/2 is simply :

$$\partial_x T(\Delta/2,t) = -Q_0 \int_0^t \frac{dt'}{\sqrt{4\pi D t'}} (1 - \exp(-\frac{\Delta^2}{4Dt'})) \quad (7)$$

This integral can be reduced, after the change of variables $t' = \theta(\Delta^2/4D)$ to :

$$\partial_x T(\Delta/2,t) = -\frac{Q_0}{D\sqrt{16\pi}} \int_0^{4Dt/\Delta^2} \frac{d\theta}{\sqrt{\theta}} (1 - \exp(-1/\theta)) \quad (8)$$

The integral above converges when its upper bound goes to infinity : $\int_0^\infty d\theta/\sqrt{\theta}(1-\exp(-1/\theta)) = 2\sqrt{\pi}$. It follows that for small values of Δ, $\partial_x T(\Delta/2,\tau) \approx Q_0\Delta/(2D)$, provided $\tau >> \Delta^2/(4D)$. As expected, the absolute value of the temperature derivative $\partial_x T(\Delta/2,t)$ is a monotonically increasing function of Δ (the larger the region where heat is released, the higher the flux). For a given value of $(\partial_x T)_c$, there exists a critical value Δ_c defined implicitly by :

$$(\partial_x T)_c = \frac{Q_0\Delta_c}{D\sqrt{16\pi}} \int_0^{4D\tau/\Delta_c^2} \frac{d\theta}{\sqrt{\theta}} (1 - \exp(-1/\theta)) \quad (9)$$

such that for $\Delta < \Delta_c$, the critical value is not reached while the Q term is on, and for $\Delta > \Delta_c$, the critical value is reached at a time t < τ. In the former case, the solution will relax for t > τ towards the value at x$\to\infty$, and propagation of a wave does not start. In the other case ($\Delta > \Delta_c$), the source term gets turned on for $t > \tau$ in the initially unfractured region, resulting in a propagating wave. Provided $\Delta_c / (D\tau)^{1/2} \ll 1$, Eq.9 reduces to :

$$\Delta_c \approx \frac{2(\partial_x T)_c D}{Q_0} = 2(D\tau)^{1/2} G \qquad (10)$$

where G was defined in Eq.5.

Eq.9 predicts that ignition by a brittle fractured zone becomes impossible when $G \geq \overline{G} = 1/\sqrt{\pi} \approx 0.564$ (the critical length becomes infinite). It is interesting to notice that for $\overline{G} \approx 0.564 \leq G \leq G_c \approx 0.65$, ignition of propagating waves by this mechanism is impossible. The heat flux generated at the boundary of the fractured zone is never large enough. Ignition of waves requires other mechanisms, such as local heat perturbations.

The predictions for the critical size have been tested numerically in the small G regime. We explicitly checked that the maximum derivative is located at the edge of the fractured zone. A linear dependence of Δ as a function of G is observed for small values of G, in complete agreement with the analysis above. Figure 1 shows the dependence of the critical nucleus as a function of the parameter G.

The analysis presented here can be extended to 2 and 3 dimensions. The results are qualitatively completely similar to the one-dimensional case. Figure 2 shows the result in 3-dimensions.

Results and discussion

The data available in the literature (1,2) allow us to estimate the value of Δ_c in real systems. In the case of the chlorination of butylchloride, in liquid nitrogen (T≈77 K), $\tau \approx 0.1s$, $Q_0 \tau \approx 60K$ and $(\partial_x T)_c \approx 300K/cm$ (1,2) . The heat diffusivity D is of the order D ≈10⁻² cm²/s (or less), so G≈ 0.15, and $\Delta_c \approx 5.10^{-3}$cm. Effectively, in the experiments, the critical length is very small. We thus obtain the physically very important result : in order to initiate propagation of a travelling wave of cryochemical transformation, it is enough to rupture locally a very small piece of the solid matrix of reagents, of size of the order of 10μm.

To check the presented theoretical results, experiments with reaction of butilchloride chlorination were carried out in liquid nitrogen (77 K) with

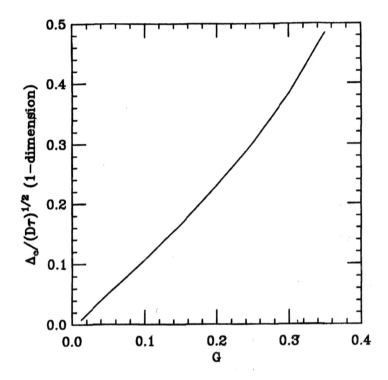

Figure 1 . The critical size of the fractured zone necessary to start a propagating wave, Δ_c, a function as of $(\partial_x T)_c$ in 1-dimension. Note that the physical quantities have been made dimensionless (the relation between G and $(\partial_x T)_c$ is specified by Eq.5).

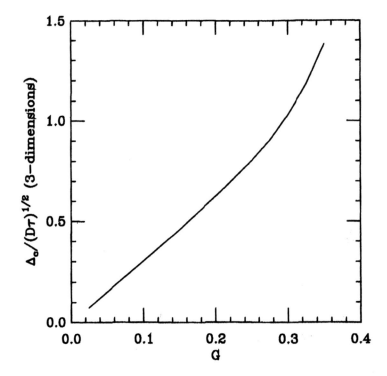

Figure 2 . The critical size of the fractured zone necessary to start a propagating wave, Δ_c, a function as of $(\partial_x T)_c$ in 3-dimension. The results are qualitatively completely similar in 1 and in 3-dimensions ; compare with Fig. 1.

molecular ratio of the reagents in initial solid mixture   3:1. The solid mixture was received a preliminary treatment by gamma-radiation. Similar experiments were carried out with polymerization reactions (polymerization of acetaldehyde, or co-polymerization of isoprene and sulfur dioxide). The two following geometries were used: (i) in a cylindrical glass vessel (diameter of ~1 cm) the needle was pressed on the upper open surfase of the sample of reagents, and (ii) in a thin film of reagents (of the order of 100 μm) frozen on the bottom of a flat dish.

In both cases prick and scratching of the samples initiated locally the reactions, which then lead to the propagation into the whole sample, in the autowave, "layer-by-layer" mode. The local disruption of the sample over a very small region (of the order of a few microns near to diameter of the needle) was sufficient action to ignite the reactions. This result confirms the prediction of the theoretical analysis.

We emphasize that only brittle disruption was capable to ignite the reaction. A fast interaction of the needle with the sample, involving time scales of order 0.01-0.1 s is necessary to induce the effect. This is reminiscent of what happens in shock waves. If the mechanical loading of the needle is much slower (times scale of the order of 1 ms), the reaction is not ignited. It means that a plastic deformation is not capable to excite the reactions, and, as suggested by theory, only elastic deformations and following brittle disruption is a necessary condition for "cold ignition" of cryochemical reactions in solids.

Conclusion

The results above, used to describe chemical reactions at very low temperature, allowed us to propose an unexpected application of the "cold ignition" phemomenon in a very different field of chemistry, namely for high temperature processes of combustion and explosion.

In spite of intensive research for many decades, the theory of combustion and explosion has not been able to explain the phenomenom of sensitivity of solid crystalline explosives to shock and friction. The phenomena is well known, and in practice often used (for the ignition of solid explosives by shock or by a short dynamical action), but is still waiting for a theoretical explanation. The thermal approach was used to address the problem: mechanical action (shock, friction) is transformed into thermal energy, and after this, the reaction develops according to the classic combustion theory, based on thermal activation. This approach did not lead to a quantative description of the phenomenon.

We argue that the theoretical picture used here to explain the "cold" ignition of solid phase reactions may help to describe the phenomenon of sensitivity of solid explosives to shock and friction. The results of experiments on ignition of explosion of azides of heavy metals (*3*) corroborate this prediction. In this sense, the experiments with "cold" ignition in cryochemical solid phase reactions provide a convenient and safe experimental model for investigating the phenomenon of sensitivity of solid explosives with shock and friction. We mention that the theoretical and experimental investigations of the phenomenon of "cold" ignition in solid phase cryochemistry allowed us to propose new mechanisms to describe the dynamics of the "gasless detonation" process (more detail can be found in (*4,5,6*).

Another domain where the ideas and concepts developed here may be useful concerns the synthesis of increasingly complex elements in a cold universe. With such a scenario in mind, thermal stresses in the solid core of a planet, or in dust particles in interstellar clouds could be the driving force in the synthesis of methane and ammoniac from simple elements, and at a later stage, in the synthesis of molecules of biological relevance, such as amines (*7*) .

This work has been supported in part by NATO under contract SfP.97 1897, and by the program PIC 1170 from CNRS.

References

1. Kiryukhin, D. P. ; Mozhaev, P. S ; Barelko, V. V., *Sov. J. Chem. Phys.* **1992**, *11*, 372.
2. Kiryukhin, D. P. ; Barkalov, I. M., *Polymer Science, Ser B*, **2001**, *42*, 244.
3. Barelko, V. V. ; Ryabykh, S. M. ; Karabukaev, K. Sh. ; *Sov. J. Chem. Phys.* **1994**, *12*, 377.
4. Pumir, A. ; Barelko, V. V. ; *Eur. Phys. J. B* , **1999**, *10*, 379.
5. Pumir, A. ; Barelko, V. V. ; *Eur. Phys. J.B* , **2000**, *16*, 137.
6. Pumir, A. ; Barelko, V. V. ; *Eur. Phys. J.B* , **2001**, *22*, 71.
7. Barelko, V. V. ; Barkalov I. M. ; Kiryukhin, D. P. ; Pumir, A. ; *Macromol. Symp.*, **2000**, *160*, 21.

Chapter 14

Evolution of Isothermal Polymerization Fronts via Laser Line Deflection and Predictive Modeling

Lydia L. Lewis[1], Cynthia A. DeBisschop[2], John A. Pojman[3], and Vladimir A. Volpert[4]

[1]Department of Chemistry, Millsaps College, Jackson, MS 39210
[2]Department of Mathematics and Statistics, Old Dominion University, Norfolk, VA 23529
[3]Department of Chemistry and Biochemistry, The University of Southern Mississippi, Hattiesburg, MS 39406
[4]Department of Engineering Sciences and Applied Mathematics, Northwestern University, Evanston, IL 60208

Isothermal frontal polymerization (IFP) is a self-sustaining, directional polymerization that can be used to produce gradient refractive index materials. Accurate detection of frontal properties has been difficult due to the concentration gradient that forms from the diffusion and subsequent polymerization of the monomer solution into the polymer seed. A laser technique that detects tiny differences in refractive indices has been modified to detect the various regions in propagating fronts. Propagation distances and gradient profiles have been determined both mathematically and experimentally at various initiator concentrations and cure temperatures for IFP systems of methyl methacrylate with poly(methyl methacrylate) seeds and with the thermal initiator 2,2'-azobisisobutryonitrile.

170

Isothermal frontal polymerization (IFP) is an autocatalytic, directional, and self-propagating polymerization. IFP occurs when a solution of monomer and thermal initiator diffuse into a piece of polymer (polymer seed) creating a viscous region that supports the Trommsdorff, or gel, effect. Because the rate of polymerization is faster in this viscous region than in the monomer solution, the currently-reacting polymer propagates from this viscous region producing a new viscous region into which more monomer solution can diffuse. This monomer diffusion couples with the gel effect to produce an autocatalytic polymerization wave that travels through the monomer solution in a similar fashion to combustion waves. The wave continues to propagate through the monomer solution until the monomer solution homogeneously polymerizes on top of the front halting the front (Figure 1)(1).

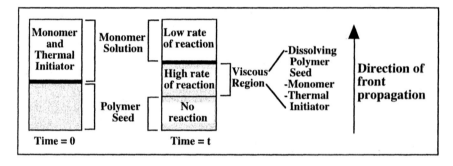

Figure 1. Mechanism of IFP

IFP can produce gradient materials when a dopant is inserted into either the seed or monomer solution(1,2). The dopant can be a second monomer that copolymerizes with the first monomer or can be a material that does not react with the monomer. The dopant material is incorporated into the polymer matrix at a different rate than the polymerizing monomer(3), forming a concentration gradient of the dopant material. This concentration gradient produces special properties within the polymer and is specific to the dopant. For example, when the dopant material has a different refractive index than the monomer, a gradient in the overall refractive index is formed, which causes light to deflect as it travels through the material. This refraction can produce a magnification effect when the gradient is along the radius of a flat lens and can produce an attenuation effect when the gradient is along the radius of a rod(4). Gradient materials are found in materials ranging from waveguides to lens arrays(1).
Extensive work has been done on gradient systems to improve their optical

properties. Typically, these materials are produced with copolymerization reactions, and the optical properties have been improved by altering the monomer ratios and/or the structures of the monomers(5). Studies of the fundamental mechanism of IFP (without dopants) includes work to determine the lifetime of the front(6) and include work using polymeric inhibitors to lengthen these lifetimes(7-9). Various mathematical models to simulate these fundamental IFP systems include models with and without polymeric inhibitors(7,10,11). Physical properties, such as the propagation distance, have been difficult to accurately determine during the course of the reaction: The reaction (even without a dopant) produces a concentration gradient, $\Delta C/\Delta y$ (change in the concentration over a distance), of the monomer solution, of the currently-reacting polymer, and of the finished polymer. All of whose boundaries are indiscernible to the naked eye (Figure 2). Light deflects as it passes through a gradient, and this fact has been used to monitor diffusing systems(12). This technique has been modified to produce a non-invasive monitoring technique, termed laser line deflection (LLD), that detects the reaction zone of propagating fronts without disturbing or altering the reaction.

IFP systems without dopants using the monomer methyl methacrylate (MMA) with poly(methyl methacrylate) (PMMA) seeds and with the thermal initiator 2,2'-azobisisobutyronitrile (AIBN) were run under various reaction conditions. The systems were monitored using LLD, which allowed propagation distances and gradient profiles to be determined over time. Fronts were run at different cure temperatures (42 to 68 °C) and using different initiator concentrations (0.03 to 0.15% AIBN to MMA (wt./V)) at each cure temperature. To better predict the properties of IFP products and to better understand the behavior of the IFP mechanism, a mathematical model of IFP based on radical polymerization and mass diffusion of the monomer species was developed(11). Propagation distances and gradient profiles were determined using the experimental parameters, and comparisons were made between the experimental and mathematical results to gauge the effectiveness of the model.

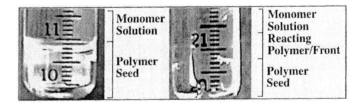

Figure 2. IFP Sample. A. Time = 0. Polymer seed and monomer solution are easily distinguishable. B. Time = 2 hours. Monomer solution and currently-reacting polymer in the reaction zone are indistinguishable.

Theory of Laser Deflection for Propagating Fronts

When light passes from one substance through another substance normal to their boundary, the light is not deflected, but its speed does change (Figure 3A). If a refractive index gradient, $\Delta\eta/\Delta y$ (change in refractive index with respect to position), is present in the y-direction of the medium, the light deflects as it passes through the substance. Figure 3B illustrates this deflection using a wave line: The refractive index gradient is smaller than the wave. Thus, one end of the wave travels through the medium faster causing the wave to deflect, or bend, and the out-going light is deflected from the position of the incident light(*13*).

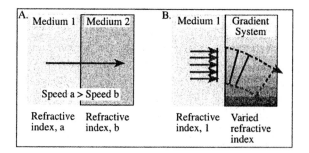

Figure 3. A. Light passing through two media of different refractive indices normal to their boundary is not deflected. B. Light passing through a medium containing a refractive index gradient is deflected.

The change in refractive index, or the refractive index gradient, $(d\eta/dy)$ is related to the change in concentration by Equation 1

$$\frac{dC}{dy} = \frac{dC}{d\eta}\frac{d\eta}{dy} \qquad (1)$$

where C is the concentration, where η is the refractive index, and where y is the y-axis. To determine the position of the reaction zone within an IFP sample (the region between the pure monomer solution and between the finished polymer where a concentration gradient exists), the light can be passed through the sample along the z-axis (Figure 4). If the out-going light does not deflect, the light passed through a pure substance, e.g. monomer solution, polymer seed, or finished polymer (Sections A and C in Figure 4). If the out-going light is deflected, the light passed through a region that contains a refractive index gradient (and a concentration gradient), e.g. newly currently-reacting polymer and finished polymer (Section B in Figure 4).

To view an entire IFP sample simultaneously and to prevent disturbing the reaction, the reaction vessel is illuminated from the bottom corner to the opposite top corner using a line laser (Figure 5). (Illumination in the vertical plane would result in portions of the deflected light being hidden by the incident light.) The gradient region exists where the light is deflected (the "dip," Figure 5). Over time, the reaction zone moves vertically up the vessel as shown by the movement of the two horizontal lines in Figure 6. The "dip" is a relative measure of the steepness of the gradient and can be mathematically transformed to show the gradient profile, $d\eta/dy$ or dC/dy. The front position is defined as the maximum change in the refractive index with respect to the y-axis, $\Delta\eta/\Delta y = 0$. Front propagation is defined as the position over time of this maximum.

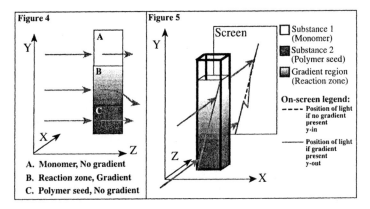

Figure 4: Light Passing Through Gradient and Non-Gradient Regions
Figure 5: LLD to View Entire IFP Sample Simulataneously

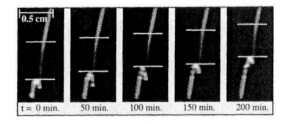

Figure 6: Time Progression of LLD Sample (0.15% AIBN in MMA at 66-68 °C).

Mathematical Model

IFP has been observed with radical polymerization. The three main reactions in the radical polymerization mechanism are 1. initiator decomposition where the initiator, I, decomposes to usually form two radicals,

D_0, with an initiator efficiency, f, 2. propagation of the polymer chain where a radical D_n reacts with monomer, and 3. termination where two polymer radicals couple to form dead polymer.

1. $\qquad\qquad\qquad$ I $\quad\rightarrow\quad$ 2 f D_0 $\qquad\qquad\qquad$ k_d

2. $\qquad\qquad\qquad$ $D_n + M \quad\rightarrow\quad D_{n+1}$ $\qquad\qquad\qquad$ k_p

3. $\qquad\qquad\qquad$ $D_n + D_m \quad\rightarrow\quad$ P $\qquad\qquad\qquad$ k_t

The reaction rate constants are given by the Arrhenius equations

$$k = k_j(T) = k_j^0 \exp(\frac{-E_j}{R_g T}) \qquad j = d,p,t$$

where k_j^0 is the frequency factor, where E_j is the activation energy, and where R_g is the ideal gas law constant.

The mathematical model includes kinetic equations based on the rate equations of initiator decomposition, radical consumption, and monomer consumption and includes mass balances of these species(*14*).

$$\frac{\partial I}{\partial t} = -k_d I$$

$$\frac{\partial D}{\partial t} = 2 f k_d I - 2 k_{tg} D^2$$

$$\frac{\partial M}{\partial t} = D_M \frac{\partial^2 M}{\partial y^2} - k_p DM$$

The model includes boundary conditions

$$M(0,t) = 0, \qquad \frac{\partial M(L,t)}{\partial y} = 0$$

and initial conditions

$$M(y,0) = M_0 H(y - y_0) \qquad I(y,0) = I_0, \qquad D(y,0) = 0$$

where y and t are the coordinates for space and time, where D_M is the monomer diffusion coefficient, and where H is the Heaviside step function. The termination reaction rate constant, k_{tg}, which characterizes the gel effect, depends on the degree of conversion. (See references 11 and 14.)

To calculate the numerical front propagation, these conditions along with the kinetic equations were nondimensionalized and were transformed to a coordinate system that moved with the front. To increase the resolution, more points were added near the front.

In order for the front to occur, the gel effect must take place and occurs when the viscosity reaches a critical value causing the rate of polymerization to dramatically increase. This point of conversion was plotted wrt time to show the propagation, and LSODA was the software used to generate the numerical solutions. The gradient profile of the monomer concentration, dM/dy, was calculated using finite differences.

Experimental

MMA (98%, Aldrich) was stored over molecular sieves prior to use, run through an inhibitor-removal column, and treated with CaH_2 to further remove water, which hindered optical clarity and, thus, front detection. AIBN (98%, Aldrich) was recrystallized in methanol. 2,2',6,6'-tetramethyl-1-piperidinyloxy (TEMPO, 99%, Aldrich) was used as is. Stock solutions of various percentages of AIBN in MMA (wt./V) were mixed and separated into 10.0 mL aliquots. All solutions were purged with breathing-quality air for ten minutes immediately prior to use. (Oxygen is a polymeric inhibitor. The air purge ensured that similar quantities of oxygen were dissolved in the samples for similar amounts of inhibition.) Polymer seeds that would fit into standard glass cuvettes (1x1x4 cm, width x pathlength x height) were produced by narrowing cuvettes to 0.9x0.9x4 cm using trimmed, glass microscope slides. These cuvettes were filled with stock solution (0.03% AIBN to MMA), which was polymerized over the fluctuating temperature range of 47 to 52°C for twenty-four hours. Slices of approximately 6.0 ± 0.5 mm were cut using a Gryphon® diamond wire saw.

To run a front, a polymer slice (or seed) was placed in the bottom of a standard glass cuvette, and monomer solution of the concentration of interest was poured on top. The cuvettes were capped with inverted polyethylene caps (Fisher Scientific) to minimize the air/liquid interface. (If an air layer were present on top of the samples, the oxygen in the air would, over time, have diffused into the upper layer of monomer solution and would have inhibited polymerization in this layer.) The cuvettes were then placed in the cuvette holder in the pre-heated thermostatic box (Figure 7), and the fronts were run at the desired temperature until completion.

Fronts of concentrations of 0.03, 0.06, and 0.15% (AIBN to MMA) were run at the temperatures ranges of 42 to 47, 47 to 52 and 66 to 68 °C. The ranges were a result of temperature fluctuations due to the type of thermostat used. Three samples at each set of conditions were run to determine an average and standard deviation. To ensure that the fronts were initiated by the polymeric seed, controls were run that did not contain seeds. To ensure that the gradient movement detected was not solely seed dissolution, controls using seeds and solutions of 4.0% TEMPO, a free-radical scavenger used as an inhibitor, in

MMA were used to observe seed dissolution. Due to cost, only one control for each set of experimental conditions was run.

The fronts were monitored via laser line deflection (LLD): A 5.0 mW laser (Mellos Griott) was tilted 13.5 degrees from the vertical axis and passed through a cylindrical lens (f = 100 mm, 60x50 mm) (also tilted) to narrow the width of the beam to approximately 0.6 mm. The laser was then passed through the sample, and the image was shone on an opaque, plastic screen. A black-and-white camera, which was connected to a computer, was placed directly behind the screen. All components were mounted on posts, post holders, and rail carriers, which were mounted on an optical rail. All components except the laser, camera, and computer were housed in a thermostatic blackout box, and the freeware program NIH Image® was used to capture images of the fronts at various time intervals (Figure 7).

Acrylates are categorized as skin irritants with MMA listed as a milder irritant. All steps involving the volatile MMA were performed in a fume hood until the samples were capped, which prevented further volatilization. PMMA is inert offering no danger. AIBN and TEMPO are listed as toxic. The quantities of AIBN and TEMPO used were minute and were handled with appropriate safety equipment to prevent any contact or inhalation. Polymer samples cut with the saw were held in a vise specifically designed to hold these small samples while cutting without fear of bodily harm. The five milliwatt laser was housed in a black-out box and angled to prevent lasing of eyes. All appropriate safety equipment, e.g. glasses and solvent-resistant gloves, were utilized.

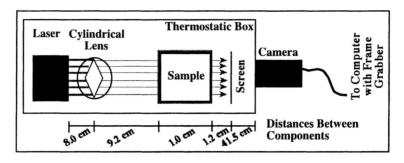

Figure 7. LLD Apparatus

Data Analysis

To determine propagation distances and gradient profiles, the captured images (e.g. Figure 8A) were assigned x- and y-data points using the software UN-SCAN-IT® (e.g. Figure 8B), and the data was then analyzed in a spreadsheet program: Each x-value contained several y-values because of the width of the LLD line. To reduce the scatter of the bell-shaped data, the width of the LLD line was reduced by excluding all but the lowest y-value for each x-value. To determine the propagation distances and gradient profiles, the appearance of the data was changed from the original image (containing the dip, Figure 8B) to a bell-shaped curve (Figure 8C): The position where the incident light would have been had no gradient been present (Y-in) was subtracted from the position of the out-going light (Y-out) and plotted with respect to Y-in (Figure 8C). Because the light would not have passed through a gradient, Y-in was always a straight line and was determined by linearly regressing the data from the monomer portion at the start time and applying the equation to the data for each time. (The dotted lines in Figure 8 represent this linear regression (or Y-in) and have been offset from the data for clarity in the figures.) Figure 8C shows an example of the time of the altered data progression ($d\eta/dy$ gradient profiles). As previously mentioned, the Y-in position corresponding to the gradient maximum, $d\eta/dy = 0$ (where y is "Y-in"), is the front position. These positions are indicated with the dotted arrows in Figure 8C. These front positions were determined with respect to time for a given initiator concentration and cure temperature. The times of each sample were zeroed using the time of the longest-running sample as the end time for each front. The propagation length was also zeroed using the end length of the longest-running sample as the end length for each front. For a given set of conditions, the propagation distances were averaged (e.g. Figure 9). Concentration gradient profiles, dC/dy, for a given set of conditions were determined by multiplying $d\eta/dy$ by $dC/d\eta = (M_0-0)/|\eta_{MMA} -\eta_{PMMA}|$ (Equation 1). (The graphs in Figure 8 were multiplied by $dC/d\eta$ to yield the dC/dy profiles, which are represented in Figure 10.)

Results and Discussion

Limitations of the LLD Apparatus

LLD successfully monitored IFP fronts of MMA systems with PMMA seeds as shown by the time progression of LLD images in Figure 6. It has been determined that this LLD apparatus has several limitations, which must be discussed to illustrate limitations within the experimental data: The first limitation involves the refractive indices of the components: The amount of deflection increases with increasing $\Delta\eta$. When the reactive and/or diffusing interface is infinitely small, $\Delta\eta/\Delta y$ is infinitely large. Thus, $\Delta\eta/\Delta y$ is off-scale with a thin and undetectable "dip" of light at early times. As time progresses, $\Delta\eta/\Delta y$ decreases because Δy increases allowing the gradient to fit within the limitations of the imaging system and allowing the gradient to be detected by the camera. With MMA and PMMA systems, the time necessary for $\Delta\eta/\Delta y$ to decrease is encompassed in the seed dissolution phase that must occur before the front begins. In systems with larger differences in refractive indices (e.g. water

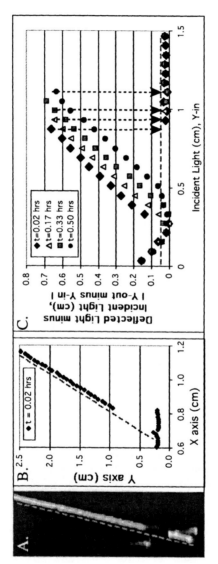

Figure 8. Progression of LLD Data Conversion. A. LLD Image B. X- and Y-Data Points from UN-SCAN-IT® C. Gradient Profile at various times. The dotted line is the Y-in for these conditions (0.15% AIBN to MMA at 66-68°C).

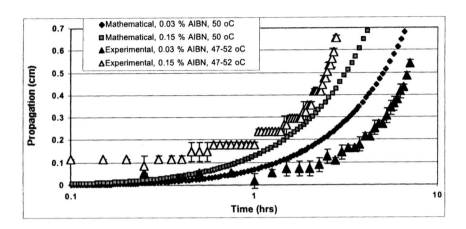

*Figure 9. Mathematical and Experimental Front Propagation at 50°C
(Mathematically) and 47-52°C (Experimentally). The Mathematical
Propagation extends off the graph to 3.02 cm in 12.24 hours for 0.15% AIBN
and to 4.64 cm in 28.6 hrs for 0.03% AIBN. The standard deviation on the
experimental data, if larger than the markers used, is shown by vertical bars.
(The velocity is the slope of the propagation.)*

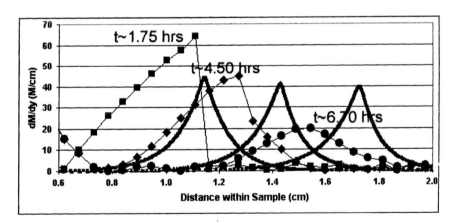

*Figure 10. Mathematical and Experimental Gradient Profiles (dC/dy) at 50 °C
(Mathematically) and 47-52 °C (Experimentally) for 0.03 % AIBN in MMA.
Curves with markers of ●, ■, and ◆ represent experimental runs. The
mathematical results are for $D_M = 2.5 \times 10^6$ cm^2/s, $f = 0.5$, $I_o = 0.03$ % (2
mmol/L), $M_o = 9.0$ mol/L, and the Arrehenius rate parameters for AIBN/MMA
system as in (11).*

and glycerin), time must be allotted for $\Delta\eta/\Delta y$ to decrease to the necessary dimensions for the imaging system.

Limitations that arise from the laws of optics include the "tangent rule" and the critical angle. The tangent rule is the fact that the projected image on the screen is magnified with an increase in the distance from the sample to the imaging screen. In order to collect all of the data from the system, the imaging screen must be near enough to the reacting system so that all of the deflected light appears on the screen and far enough away that the image is magnified to the desired size. The optimum distance from the front of the cuvette to the imaging screen for IFP systems of MMA with PMMA seeds is two centimeters. The critical angle limitation is the maximum angle of deflection where total internal reflection occurs preventing any light from appearing on the imaging screen. For the systems studied in this paper, the maximum angle of deflection was approximately twenty-four degrees, and the critical angle for total internal reflection was approximately forty-two degrees.

Discussion of the Experimental and Mathematical Data

For an increase in initiator concentration and/or cure temperature, the rate of polymerization for the mathematical and experimental data increased, which is in agreement with radical polymerization kinetics. The increase in polymerization rate shortened the time before the monomer solution bulk polymerized on top of the front stopping the front. Thus, the total propagation time was less. The mathematical propagation times ranged from seventy hours (0.03% AIBN and 40 $^\circ$C) to three hours (0.15% AIBN and 67 $^\circ$C), and the experimental propagation times ranged from approximately nineteen hours to less than one hour (0.03% AIBN at 42 to 47 $^\circ$C to 0.15% AIBN at 66 to 68 $^\circ$C). The mathematical results predicted that the propagation length would be less with this shortened propagation time, e.g. 7.4 to 1.4 cm (0.03% AIBN at 40 $^\circ$C to 0.15% AIBN at 67 $^\circ$C). The experimental results showed that the propagation length remained the same, 0.55 $\pm$0.11 cm, within the standard deviation regardless of experimental conditions except those of 0.03 % AIBN in MMA at 42 to 47 $^\circ$C, which propagated an average of 1.03 cm $\pm$ 0.08 cm. The increase in polymerization rate (from an increase in initiator concentration and/or cure temperature) also corresponds to an increase in the front velocity (the slope of the propagation curves) for both the mathematical and experimental data, which is also in agreement with radical polymerization kinetics. For any given set of reaction conditions, the mathematical and experimental velocity increases with time. Figure 9 shows representative data for the two different initiator concentrations at the same temperature for both the mathematical and experimental propagations.

The mathematical prediction that the propagation length shortens with an increase in initiator concentration or cure temperature is explained by the fact that the polymerization time of the monomer solution competes against the front velocity. The model predicted that the front velocity would not increase as rapidly as the polymerization of the monomer solution. Experimentally, the polymerization of the monomer solution, which diffuses into the polymerizing front, shortens the front polymerization time because the diffusing (and polymerizing) monomer approaches the critical concentration needed to evoke the gel effect and, thus, the front. The increased viscosity of the monomer solution is also the driving factor for the increase in velocity over time that is experienced for any given set of reaction conditions.

Figure 10 shows representative gradient profiles for both the mathematical and experimental data. For the tested reaction conditions, both the mathematical and experimental data show progression of the gradient profiles in the y-direction, which show the movement of the fronts, and both types of data show the decrease and eventual smoothing of the maxima of their profiles. This decrease is in agreement with the fact that the monomer solution homogeneously polymerizes on top of the front stopping the front. Thus, the refractive index of the monomer solution increases with time until it reaches the refractive index of the currently-reacting polymer and of the finished polymer showing that the entire contents of the reaction vessel are of a homogeneous refractive index (or concentration).

The mathematical data shows a symmetrical and narrow gradient profile indicative of a narrow reaction zone where the rate of diffusion of the monomer solution into currently-reacting polymer is the same as the rate of the currently-reacting polymer diffusing into the monomer solution. The experimental data shows an asymmetrical and wider gradient profile, which indicates that the reaction zone is wider than predicted and which indicates that the rate of diffusion of the monomer into the reacting polymer is faster than the rate of diffusion of the currently-reacting polymer into the monomer.

The experimental controls where no seeds were included polymerized homogeneously indicating that it is necessary to include a seed for IFP to occur. Controls that included seeds and a polymeric inhibitor to illustrate seed dissolution showed an initial sharp "dip" that did not move up the reaction vessel but instead diffused with time indicating no reaction taking place.

Conclusions

Examination of IFP systems with LLD shows that IFP of MMA systems with PMMA seeds and the initiator AIBN does exist and provides the most accurate and most systematic data of IFP systems to date. LLD allowed the determination of differences in refractive index over distances smaller than one millimeter to be detected in order to obtain more accurate data.

Mathematical and experimental propagation distances and times reveal qualitative trends where an increase in initiator concentration and/or cure temperature causes an increase in the propagation and velocity. Experimental propagation is on the order of 0.55 ±0.11 cm, except for 0.03 % AIBN in MMA at 42 to 47 °C (1.03 ± 0.08 cm), and ranges from less than one hour to nineteen hours for systems of 0.15% AIBN in MMA at 66 to 68 °C to systems of 0.03%

AIBN in MMA at 42 to 47 °C. Mathematically, the propagation ranges from 1.4 cm to 7.4 cm for systems of 0.15% AIBN in MMA at 67 °C to systems of 0.03% AIBN in MMA at 40 °C. Both the mathematical and experimental velocities (the slopes of the propagation curves) increase during the course of the reaction regardless of reaction conditions. This fact is in agreement with predicted results that the monomer solution polymerizes and increases in viscosity with time shortening the necessary time for the diffused monomer in the reaction zone to reach the desired viscosity for the gel effect, and thus the front, to occur.

The experimental and mathematical gradient profiles reveal qualitative trends within the systems: The gradient profiles show that the gradient maxima ($d\eta/dy = 0$) propagate with time indicating that the front does progress up the reaction vessel. These profiles also decrease over time indicating that the monomer solution does polymerize on top of the front with the entire system reaching a similar refractive index and thus, concentration. The mathematical model predicts a symmetrical and narrow reaction zone. The experimental results show a slightly asymmetrical reaction zone that has a sharp demarcation at its boundary with the monomer solution and show a wider reaction zone.

The differences between the mathematical and experimental results can result from the fact that various parameters of the mathematical model (e.g. the monomer diffusion coefficient) do not change with time as they do over the course of the experiment. Also, various parameters such as the monomer diffusion coefficient and the initiator efficiency, f, are obtained from literature where the experimental conditions, e.g. the solvent and/or cure temperature, are not the same as our experimental conditions.

Acknowledgements

The authors wish to thank their respective colleges and universities, the NASA Materials Science Program (Grant NAG8-1466), and the National Science Foundation (Grants CTS-0138660 and CTS-0138712) for funding.

References

1. Koike, Y.; Takezawa, Y.; Ohtsuka, Y. *Applied Optics* **1988**, *27*, 486-491.
2. Koike, Y.; Nihei, E.; Tanio, N.; Ohtsuka, Y. *Applied Optics* **1990**, *29*, 2686-2691.
3. Schult, D. A.; Spade, C. A.; Volpert, V. A. *Applied Mathematics Letters* **2002**, *15*, 749-754.
4. Koike, Y. Graded Index Materials and Components: in *Polymers for Lightwave and Integrated Optics: Technology and Applications*; Hornak, L. A., Ed.; Marcel Dekker, Inc.: Morgantown, WV, 1992; pp 71-102.
5. Ilyashenko, V. *Optical Fiber Communication* **1998**,
6. Golubev, V. B.; Gromov, D. G.; Korolev, B. A. *Journal of Applied Polymer Science* **1992**, *46*, 1501-1502.

7. Ivanov, V. V.; Stegno, E. V. *Polymer Science B* **1995**, *37*, 50-52.
8. Ivanov, V. V.; Stegno, E. V.; Pushchaeva, L. M. *Chemical Physics Reports* **1997**, *16*, 947-951.
9. Smirnov, B. R.; Min'ko, S. S.; Lusinov, I. A.; Sidorenko, A. A.; Stegno, E. V.; Ivanov, V. V. *Polymer Science* **1993**, *35*, 423.
10. Gromov, D. G.; Frish, H. L. *Journal of Applied Polymer Science* **1992**, *46*, 1499-1500.
11. Spade, C. A.; Volpert, V. A. *Macromolecular Theory and Simulations* **2000**, *9*, 26-46.
12. Petitjeans, P.; Maxworthy, T. *J. Fluid Mech.* **1996**, *326*, 37-56.
13. Meyer-Arendt, J. R. *Introduction to Classical and Modern Optics*; Prentice Hall: Englewood Cliffs, NJ, 1995.
14. Spade, C. A.; Volpert, V. A. *Mathematical and Computer Modelling* **1999**, *30*, 67-73.

Interfacial Systems

Chapter 15

Pattern Formation by Rim Instability in Dewetting Polymer Thin Films

Yuji Asano[1], Akitaka Hoshino[1], Hideki Miyaji[1],
Yoshihisa Miyamoto[2], and Koji Fukao[3]

[1]Department of Physics, Graduate School of Science, Kyoto University,
Kyoto 606–8502, Japan
[2]Department of Fundamental Science, Faculty of Integrated Human
Studies, Kyoto University, Kyoto 606—8501, Japan
[3]Department of Polymer Science and Engineering, Kyoto Institute
of Technology, Kyoto 606–8585, Japan

The dewetting process of thin polystyrene films from 5 to
50nm in thickness on a silicon substrate is investigated. A
spreading dry patch formed on dewetting is surrounded with a
circular rim, which stores the liquid matter expelled from the
dewetted region. When the diameter of the patch exceeds a
thickness-dependent critical value, the rim is deformed by
morphological instability; the instability gives rise to the
diversity of self-organization processes of liquid ridges or
droplets formed eventually on partially wettable substrates. A
morphological phase diagram of the deformed rim is given in
the diameter of dry patch and the initial film thickness. In the
final stage of dewetting, liquid droplets arrange to form a
"polygon network" in thick films and "polygons with random
droplets" in thin films. The origin of the instability is discussed
in terms of surface free energy.

The growing technological importance of dewetting ranges from coating of nano-scale devices to industrial lubricants, where homogeneity of liquid films is usually demanded for practical applications. Furthermore, self-organization of liquid droplets via dewetting processes nowadays attracts many researchers' interests because of the diversity of patterns formed via complicated dissipative processes coupling with each other.

When the thickness of a liquid film on a non-wettable substrate is thinner than that in equilibrium (usually ~mm), the film ruptures with cylindrical holes and shrinks from the contact line, on which three phases (liquid, substrate and air) meet. For relatively thicker films, initial holes are nucleated by dust particles and defects on the substrates. In the case of sub-micron thick films, another origin of spinodal decomposition dominates for the formation of initial holes, and hence the density distribution of the holes depends on the initial film thickness (1).

Once the cylindrical holes are created on the film, dewetting proceeds and the materials removed from the dry regions are stored into the rims of the wet region. In particular on a highly non-wettable or slippery substrate, deformation of rims by the morphological instability has been reported (2, 3). The rim instability is expected to control the dewetting dynamics involving dewetting velocity and pattern formation process in partially wetting region. In this paper, we investigate the rim instability and its effects on the arrangement of droplets formed in the final dewetting process.

Film Rupturing Induced by Spinodal Instability

Brochard and Dalliant have indicated that liquid films on non-wettable substrates are spinodally unstable against thickness fluctuation if the thickness e is thinner than 100nm (1). Their model of linearized capillary wave instability predicts that the thickness fluctuation with a characteristic wave vector q_M is amplified most rapidly to nucleate initial dry spots separated by $2\pi/q_M$ from each other. Their result is

$$q_M = \sqrt{\frac{3}{2}\frac{a}{e^2}} \quad (1)$$

where a is the molecular length defined by $a^2=|A|/6\pi\gamma$, $A=A_{SL}-A_{LL}$ is the difference between the solid-liquid and liquid-liquid Hamaker constants, and γ is the interfacial energy of the liquid-air.

Reiter has examined the rupturing properties of thin films of polymers with relatively high molecular weight and observed that the number density of holes,

N_P depends upon initial film thickness as $N_P \sim e^{-2}$, agreeing with eq (1) (2). By using a 2D fast Fourier transform of AFM topographical images of surface undulations, Xie et al. have observed more directly the properties of spinodal instability in dewetting; the thickness fluctuations with wave vector $q<q_c$ are amplified exponentially, while those with $q>q_c$ are suppressed with time (4).

Experimental

The molecular weight of atactic polystyrene used in this study is M_w=36,000 (M_w/M_n=1.06). The glass transition temperature T_g is 90°C. Polystyrene thin films were prepared by spin coating of toluene solutions on native-oxidized silicon wafers. In order to remove the solvent from the films, these films were dried at 100°C (just above T_g) in a vacuum oven for 24 hrs. The film thickness was measured by atomic force microscopy (AFM, SHIMADZU SPM-9500J) at room temperature to be from 5 to 50nm. By AFM the contact angle in equilibrium, which is a measure of wettability, was determined: θ_e=23°±1°. After holding the samples in the vacuum oven at a dewetting temperature of 180°C for appropriate time, we quenched the dewetting films at room temperature. For optical microscopy, we used monochromatic light of 546nm in wavelength. Small undulations of the deformed rims were thereby recognized as the contrast of interference fringes. We also used a hot stage (MettlerFP800) for *in situ* observation of the dewetting process at 180°C.

Results

Node creation by rim instability

Figure 1 shows a series of time evolution of a dry patch with a circular rim in a 29nm thick film at 180°C. Node-like deformations are observed in the rim with increasing diameter, and the developed nodes were slowed down to give rise to finger-like patterns. Reiter has suggested that the fingering patterns are due to interfacial slippage of the polymer films on the substrates (5) since the dewetting velocity dependent on the width of the rims, which is responsible for retardation of thicker parts of rims, is characteristic of slipping rims (6).

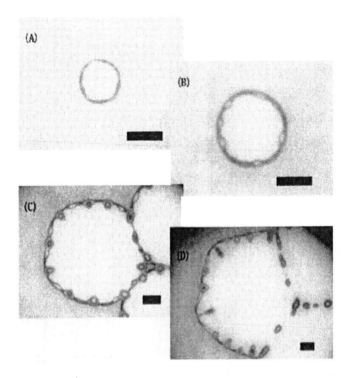

Figure 1. Optical micrographs of time evolution of deformed circular rims with nodes. The growing nodes form a fingering pattern. Scale bar is 10μm.

In order to investigate the dependence of the morphological instability on the initial film thickness, we observed the rims of dry patches at a given diameter for various film thickness: a uniform rim on a relatively thick film (Figure 2 A), a slightly deformed one with "nodes" (B), a rim with large nodes which do not catch up with the whole retreat of the rim to cause "fingering" (C) and the ensemble of droplets distributed after the fingering (D). Since the diameters of the patches are almost the same, the morphological difference is attributed to the difference in thickness of the films. We note two important aspects. Firstly not only the width of the rims but also the distance between the nodes increased with film thickness. The dependence of the wavelength of the undulations on the width of the rim reminds us of morphological instability of clamped doughnut geometry (7). Secondly, the critical diameter of the patch, at

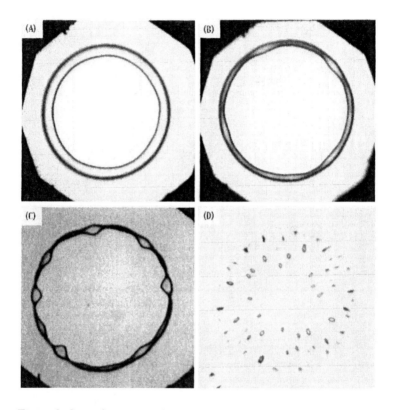

Figure 2. Optical micrographs of a dry patch in a thin film with various thicknesses. (A) 45nm, (B) 23nm, (C) 18nm, (D) 6nm. The size of micrograph is 40μm x40μm.

which the nodes become observable, also increases with film thickness. We will show later that the critical diameter and the distribution of initial holes couple with each other and determine the final dewetting patterns formed by the droplets.

Figure 3 shows the node density, which is the number of nodes per unit length of a rim, as a function of diameter of a dry patch for an 18nm thick film. The broken curve is a hyperbola corresponding to the number of nodes fixed at 6. For diameters of dry patch less than about 15μm, the number of nodes was observed to be nearly constant during the growth of the patch; for small diameter of the dry patch the node density decreases in proportion to the inverse of the diameter of the patch. When the diameter of the patch increased, new nodes appeared between neighboring nodes so that the node density remained constant. The average number of initial nodes and the constant node density are 6 and 0.12/μm, respectively for the 18nm thick film as shown in Figure 3. Both the values decreased with increasing initial film thickness (Figure 4).

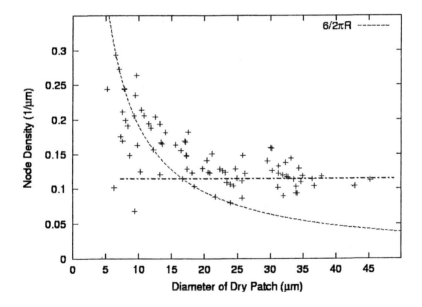

Figure 3. Node density vs. diameter of a dry patch of an 18nm thick film. The horizontal dash-dotted line is constant node density of 0.12/μm for the large diameter of dry patches.

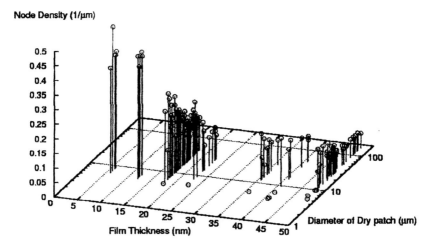

Figure 4. *Density of nodes in the circular rims around the dry patches. Uniform rims without node were observed for small dry patches of relatively thick films.*

Final droplet pattern formed by rim instability

The left side of Figure 5 (regions (a)-(c)) is a morphological phase diagram of rims in the phase space of the initial thickness of the film and the dry patch diameter; three characteristic morphologies of the dewetting rims are drawn: uniform rims (region (a) in Figure 5), deformed rims with nodes (region (b)) and rims deformed severely with the fingering (region (c)). The lateral axis (diameter of dry patch) corresponds to the time elapsed during dewetting. In the right side of the diagram (regions (d) and (e)), the final droplet patterns (see Figure 6) are drawn schematically. These patterns are formed when neighboring rims merge to be broken into droplets arranged in a pattern similar to the Voronoi polygons. For example, a dry patch in a 20nm thick film had a uniform rim with a small diameter at first, and proceeded laterally on black region in Figure 5. When the diameter of the patch reached about 10μm, the node-like deformation on the rim became observable. After that, the patch with nodes developed and eventually met neighboring patches at the diameter of 100μm. As a result, a polygon network of liquid droplets appeared in region (d). On the other hand, no uniform

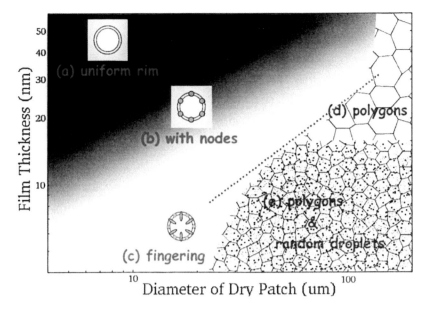

Figure 5. *Morphological phase diagram of dewetting rims (left side) and schematic sketch of the final arrangement of droplets (right side).*

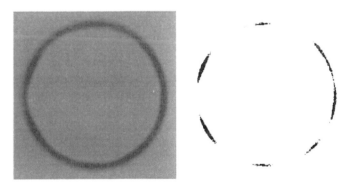

Figure 8. *Optical micrograph of a slightly deformed rim with 7 nodes (left), and its processed image to demonstrate the deformation (right). The film thickness is 18nm, and diameter of the doughnuts is 14μm.*

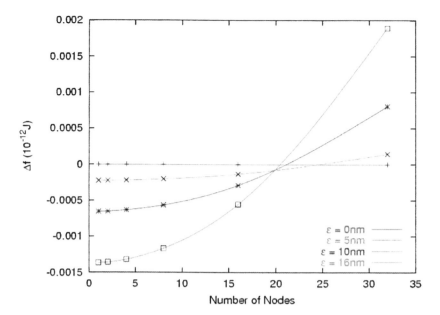

Figure 9. Computation of the difference of the surface free energies, Δf, of a uniform rim and a deformed rim. The vertical axis is the difference of the surface free energy between before and after the deformation, Δf, and the lateral axis is the number of nodes in the model rims. Three curves in the graph are for three different amplitudes of fluctuation, ε, given to thickness of the model rim, and the horizontal line corresponds to the rim without deformation, ε=0.

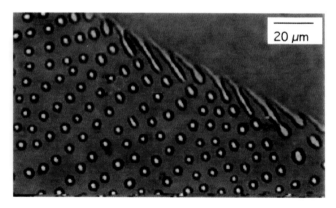

Figure 1. Optical micrograph of a receding edge of a chloroform solution of polystyrene. The solution is in the upper right corner, wheras the dewetted part is the lower left part of the picture. The receding speed of the three phase line is approx. 200 μm/s. The picture was taken in the reflection mode, since silicon was used as substrate.

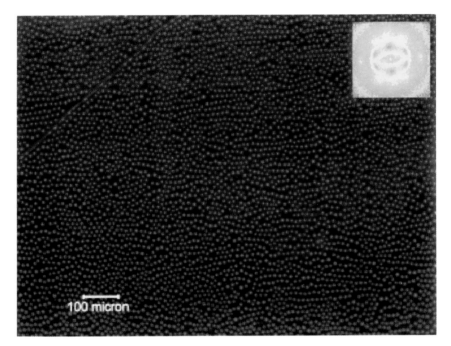

Figure 4. Fluorescence micrograph of a dewetted sample containing 4 wt% of NK85 in polystyrene. The excitation wavelength is 440-480 nm.The inset is the Fast Fourier Transformation of the largest square of the fluorescence image.

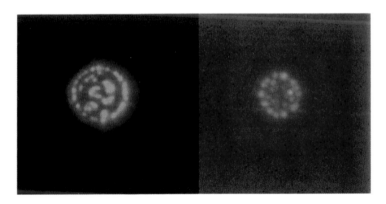

Figure 5. Fluorescence micrographs of polystyrene domes containing 4 wt% of NK85 (left) and PIC (right). The dome diameter is 6 μm in both cases.

Figure 6. Fluorescence micrographs of a polystyrene dome containing 0.1% of NK77, which forms a single fluorescing aggregate. The domediameter is 12 μm.

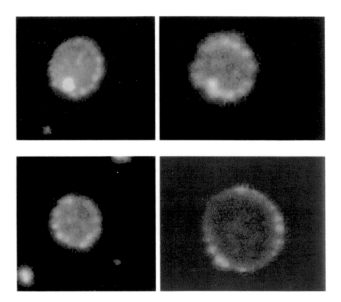

Figure 7. Fluorescence micrographs of domes where the self-organization of cyanine dyes occurred at the dome edge, leading to hierarchic structures. The diameter of the domes is in the range of 5-12 μm.

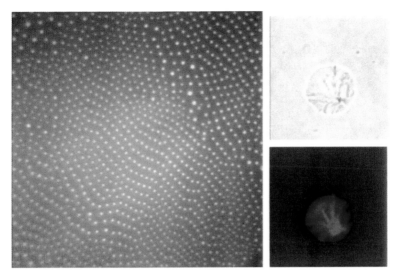

*Figure 9. Micrographs of **187** in polystyrene (50 wt%). Monochrome fluorescence image of the pattern (left) and close up of a single dome (right), in transmission and fluorescence mode. The diameter of the domes are approx. 7 μm.*

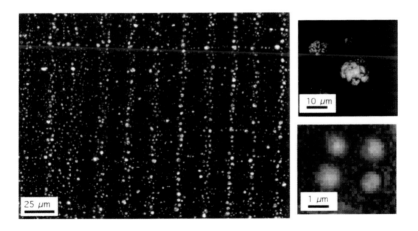

*Figure 10. Polarizing micrograph of a dewetted sample of **185** on a glass substrate. Overview of the pattern (left). Polycrystalline dome prepared by annealing at room temperature (top right) and single crystalline domes prepared by annealing at 50°C (bottom right).*

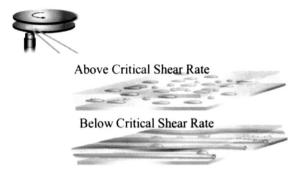

Above Critical Shear Rate

Below Critical Shear Rate

Figure 1. The droplet-string transition, viewed through a microscope. When an emulsion is sheared between parallel platens, droplets of the suspended phase (top) coalesce to form strings (bottom), with decreasing shear rate.

rim is observed in 6nm thick film. Accordingly the dry patch grew with nodes even in the early stage of dewetting. In this case, fingering pattern occurred at 6μm in diameter, and the neighboring fingering rims met each other at the diameter of 25μm. Elongated fingers were broken into several small droplets, which were enclosed by each polygon as drawn in region (e) of Figure 5. Obviously, the averaged diameter of the polygon, D_p, at a given film thickness in regions (d) and (e), is determined by the value at the corresponding boundary between region (b) or (c) and region (d) or (e). D_p has been reported to be proportional to the square of the film thickness, e, which agrees with the result predicted on the basis of the spinodal instability: eq (1) (2). We confirmed this relation in the narrow region of film thickness from 11 to 21nm; the slope of the boundary line between region (c) and region (d) or (e) for 11nm<e<21nm (green dotted line in Figure 5) satisfies the relation D_p~e^2. The upper limit would be determined by inhomogeneous nucleation due to dust particles and defects of the substrate. In a film thinner than 11nm, the uniform rim is no longer formed but the fingering region (white region in Figure 5) dominates, and hence the relation, D_p~e^2, holds in the narrow range of the film thickness.

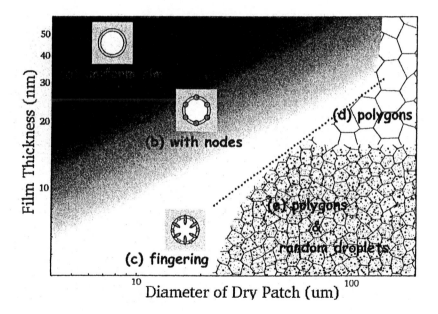

Figure 5. Morphological phase diagram of dewetting rims (left side) and schematic sketch of the final arrangement of droplets (right side). (See first page in color insert.)

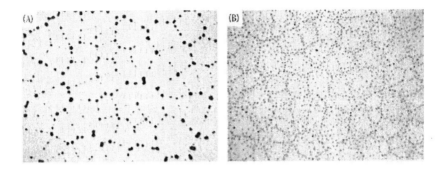

Figure 6. Optical micrographs of the final dewetting patterns at 180°C; (A): polygons for a 29nm thick film, (B): polygons and random droplets for a 6nm thick film. The width of micrograph is 730μm (A), 186μm (B) respectively.

Discussion

In order to understand the origin of the rim instability, we evaluated the surface free energy of a static rim (Figure 7). We ignored the energy contribution of disjoining pressure and gravity, which have played important roles in spinodal instability (*1*) and morphological instability (*7*). In fact they influenced little on the present evaluations; this means that our model rims were so thick that the former hardly contribute to the energy change in the deformation describe below and thin enough to be free from the latter. Figure 7(A) is a model of a uniform rim without node. We gave sinusoidal deformation to the width of the rim under the condition of constant volume. Figure 7(B) is a model rim with 7 nodes, which corresponds to a rim observed in a film with the thickness of 18nm (Figure 8). Figure 9 shows the surface free energy calculated for a deformed rim as a function of the number of nodes. With increasing deformation amplitude, ε, the slope of the curves becomes steeper with the number of nodes. It is to be noted that there exists the region where the surface free energy difference Δf is negative for deformation with small number of nodes. In that region spontaneous deformation can occur. Note that the critical number of nodes for the spontaneous deformation is about 20. Above the critical number, the surface free energy difference is positive and increases with the number of nodes and hence the rim is stable against fluctuation.

In the light of the surface free energy (Figure 9), the most stable rim has just one node. But the number of nodes realized is the number with the fastest growth rate, and hence is determined not only by the thermodynamic surface free energy but also by a dissipative process of polymer liquid in the rim. Vrij et al. have given a simple explanation in their study on the rupture of soap films: if the wave number of the fluctuation is small, the growth rate of the fluctuation will be suppressed since the liquid has to be transported over long distance under the viscous resistance against flow (8). Brochard et al. has calculated a growth rate of a fluctuation for straight ridge geometry under influence of gravity (9). By analogy, we can estimate that the wave number of the fastest mode will be about 60-70% of critical value, agreeing fairly well with the experiment.

However, in the case of the moving rims, kinetic factors must be considered such as time-dependent critical number of nodes and competition between the dewetting velocity and the growth rate of the fluctuation. The critical node number increases with the initial diameter of the dry patch (Figure 10). Consequently, the wave number of the fluctuation that grows most rapidly will increase with time. When a deformation mode can develop to be observable, the time required for the growth of the fluctuation mode has to be much shorter than the time scale of expansion of the dry patch.

When the sinusoidal fluctuation develops into the node-like deformation as shown in Figure 2(C), the above discussions will be no longer valid. In this regime, new fluctuation appears on the rims clipped by the nodes, and investigations on the properties of rim instability, e.g. node density, become more difficult since the width of the rims will be decreased due to flow into the nodes or fingers. However, this complicated, and presumably non-linear physical mechanism is interesting because auto-optimization of dewetting velocity begins in this regime (4).

Conclusions

The dewetting rim is deformed by the morphological instability. The wave number of the deformation and the critical diameter of the dry patch depend upon the initial film thickness. The rim instability and the origin of the initial holes (e.g. spinodal surface instability) determine the final arrangement of the droplets. From the calculation of the surface free energy of deformed rims, we showed that the rim instability is due to the three-dimensional structure of the rims. However, estimation of the node density is complicated by the two kinds of kinetics, the expansion rate of the dry patches and the growth rate of the rim fluctuations.

(A)

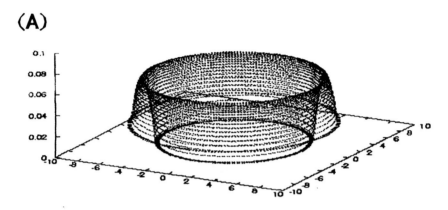

(B)

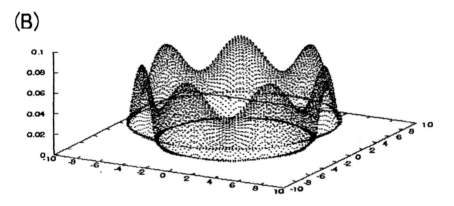

Figure 7. Models of a uniform rim (A) and a deformed rim with 7 nodes (B). Film thickness is 18nm, diameter of the doughnuts is 14µm, and the contact angle is 7°.

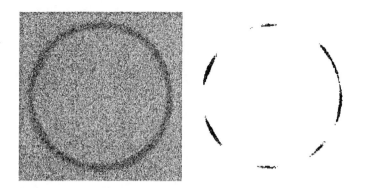

Figure 8. Optical micrograph of a slightly deformed rim with 7 nodes (left), and its processed image to demonstrate the deformation (right). The film thickness is 18nm, and diameter of the doughnuts is 14μm.
(See first page in color insert.)

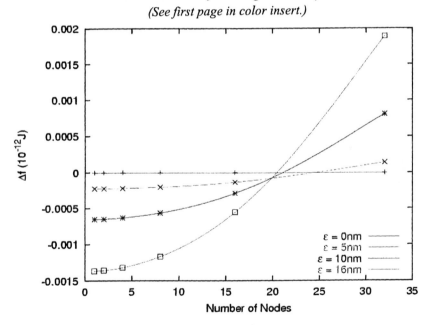

Figure 9. Computation of the difference of the surface free energies, Δf, of a uniform rim and a deformed rim. The vertical axis is the difference of the surface free energy between before and after the deformation, Δf, and the lateral axis is the number of nodes in the model rims. Three curves in the graph are for three different amplitudes of fluctuation, ε, given to thickness of the model rim, and the horizontal line corresponds to the rim without deformation, ε=0.
(See second page in color insert.)

198

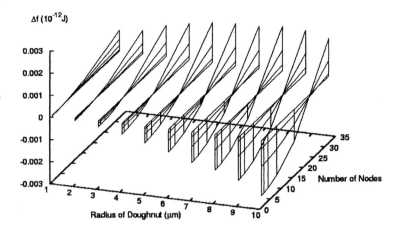

Figure 10. Surface free energy difference Δf for various radii of dry patches.

References

1. Brochard-Wyart, F.; Daillant, J. *Can. J. Phys.* **1990**, 68, 1084-1088.
2. Reiter, G. *Phys. Rev. Lett.* **1992**, 68, 75-78.
3. Reiter, G. *Langmuir* **1993**, 9, 1344-1351.
4. Xie, R.; Karim, A.; Douglas, J. F.; Han, C. C.; Weiss, R. A. *Phys. Rev. Lett.* **1998**, 81, 1251-1254.
5. Reiter, G.; Sharma, A. *Phys. Rev. Lett.* **2001**, 87, 166103.
6. Redon, C.; Brzoska, J. B.; Brochard-Wyart, F. *Macromolecules* **1994**, 27, 468-471.
7. Sekimoto, K; Oguma, R.; Kawasaki, K. *Ann. Phys.* **1987**, 176, 359-392.
8. Vrij, A.; Overbeek, J. Th. G. *J. Am. Chem. Soc.* **1968**, 90, 3074-3078.
9. Brochard-Wyart, F.; Redon, C. *Langmuir* **1992**, 8, 2324-2329.

Chapter 16

Hierarchic Mesoscopic Patterns of Dye Aggregates in Self-Organized Dewetted Films

Olaf Karthaus[1,3], Kazuaki Kaga[2], Junji Sato[1], Shigeya Kurimura[1], Kiyoshi Okamoto[1], and Toshiro Imai[1]

[1]Department of Photonics Materials Science and [2]Graduate School of Photonics Science, Chitose Institute of Science and Technology, Chitose, Japan
[3]PRESTO, Japan Science and Technology Corporation, Tokyo, Japan

By using a dewetting process micrometer-sized "domes" of polymers were formed on solid substrates, and it was found that the aggregation state and thus the photophysical properties of incorporated dye molecules depend on the dome size. During dewetting two processes compete, the formation of polymer domes and the aggregation of the dye. Thus, due to the dynamics of the process self-organized mesoscopic and hierarchic structures can be formed.

199

Dewetting is a process in which a metastable film on a substrate ruptures and forms holes, strings or droplets (*1-7*). Wetting and dewetting is important for almost all applications where thin films are used. Lubrication, adhesion, and photolithography are, among others, processes which depend on the control of the substrate-film interaction. Protective coatings and super-repellent fractal surfaces are two applications for which complete wetting and perfect dewetting is desired, respectively. For almost all applications in electronics and photonics, dewetting is an unwanted process, since it destroys the homogeneity of the thin film. Liquid crystal displays and organic light emitting diodes are just two examples where the device performance depends crucially on the formation of defect-free thin films of polymeric or low-molar-mass compounds. Thus there is a long standing and keen interest in understanding dewetting and it has been extensively studied, both theoretically (*3*) as well as experimentally (*1, 2, 4-7*). Most of the work on the dewetting of polymers focused on the dewetting of an initial metastable film. Below the glass transition (Tg) of the polymer, the film is stable. Upon heating above Tg, dewetting takes place (*1, 2*). Holes are formed at random positions in the films. These holes grow with time and may form Voronoi patterns of connected polymer strings and then finally form droplets on the substrate. Since the process is spatially and temporally random, the resulting droplets may have a broad size distribution and random spatial location without any order.

Another method to form dewetted films is the casting of a dilute polymer solution onto solid substrates (*8*). In case of volatile solvents, the volume of the polymer solution decreases with the beginning solvent evaporation. Thus the diameter of the solution drop becomes smaller, and the edge of the solution recedes towards the center of the drop. A so-called fingering instability develops and micrometer-sized polymer-containing solution droplets remain on the substrate. In the case for an undisturbed receding and dewetting, regular arrays of round polymer aggregates, or 'domes' are formed. The origin of this pattern formation lies in the complex nonlinear and dynamic processes that occur in an evaporating solution (*9*). After the solution was placed onto the substrate, solvent evaporation starts. The negative heat of evaporation leads to the cooling of the droplet surface and thus to convection within the droplet. Furthermore, evaporation of the solvent leads to an increase in solute concentration, which is most pronounced at the droplet edge, due to the lens shape of the droplet. These two effects, convection and concentration gradiation, leads to spatial inhomogeneous concentration of the solute and thus to solute-rich and solute-poor areas. Since the solute often is surface active, this difference in concentration implies a spatial difference in surface tension. Thus Marangoni convection (*10*) may lead to a growth and stabilization of the concentration gradients, especially at the droplet edge. This leads to the observed fingering instability with a regular spatial interval along the solution edge (*11*). Since the solvent evaporation is continuous, the droplet becomes smaller and the droplet edge will eventually recede. Thus the solute-rich fingers will be elongated and

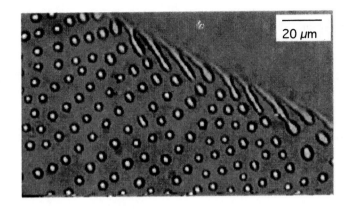

Figure 1. Optical micrograph of a receding edge of a chloroform solution of polystyrene. The solution is in the upper right corner, wheras the dewetted part is the lower left part of the picture. The receding speed of the three phase line is approx. 200 μm/s. The picture was taken in the reflection mode, since silicon was used as substrate.
(See second page in color insert.)

finally solute-rich microdroplets will remain isolated on the substrate (*8*) (Figure 1).

The pattern formation process depends very much on the substrate/solution interaction. Good results can be achieved by using hydrophilic substrates, such as mica, glass, or silicon, and hydrophobic solvents and polymers, such as chloroform and polystyrene, respectively. By using one of these combinations, dewetting takes place, since the total surface energy of the homogeniously covered substrate is higher than the dewetted substrate. The dome size can be controlled by the evaporation speed of the solvent and the polymer concentration. Faster speed and lower concentraton lead to smaller domes. Ordered arrays of domes with a diameter ranging between 200 nm and 10 μm have been realized. Interestingly, the spacing of the domes is always in the range of 3-7 times the dome diameter. After the evaporation of the solvent, the domes are solidified, and thus the pattern does not change over time.

Figure 2 shows scanning electron microscope (SEM) pictures of a polystyrene pattern. Clearly the dome shape of the micrometer-sized polymer droplets can be seen. Atomic force microscopy allowed the measurement of the local contact angle of the domes. It turned out that it is in the range of 5-8° for polystyrene domes on mica (*8*), which is slightly smaller than the equilibrium contact angle of 10°.

It could be shown that many different polymers can form dewetted microscopic dome and line structures. DNA can act as a conducting molecular wire and mesoscopic dewetted line patterns of DNA (*12*) and photoconducting poly(hexyl)thiophene (*13*) might be useful as low-dimensional electric conductors. In addition to these chain-like macromolecules, also polymers with different geometry were shown to form dome patterns. One example are dendrimers which contain a photochromic diarylethene core, which might be useful for optical data storage (*14*). In these examples the polymers, which form the dewetted pattern itself, acts as the functional materials. Since patterns are generally formed only from amorphous materials, such as polymers, the range of functional materials that can be patterned is somewhat limited.

In order to expand the applicability of mesoscopic dot-, line- and honeycomb patterns in the field of photonics and electronics, one promising approach is the incorporation of photofunctional low-molar-mass compounds into the dewetted polymer patterns, or the pattern formation of low molar mass compounds themselves.

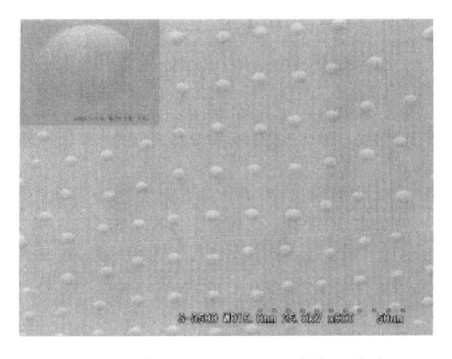

Figure 2. Scanning electron microscope pictures of a dewetted polystyrene sample on mica. The inset is a detailed view of a single dome. The diameter of the domes is in the range of 2 μm.

Dye J-Aggregates

The so-called J-aggregates of cyanine dyes are known since the 1930s. Jelly (*15*) and Scheibe (*16*) independently discovered that ionic dyes of the cyanine type form aggregates in aqueous solution. They are characterized by a very narrow and red-shifted electronic absorption peak, compared to that of the molecularly dispersed state. The fluorescence quantum yield of J-aggregates is very high, and thus J-aggregates show a bright fluorescence with a short lifetime of the excited state. Since the dye molecules are closely packed in the aggregates, the vibronic degrees of freedom of the dye molecules are limited, and the the sharp fluorescence spectrum shows only a small Stoke's shift (*17*). All these qualities, sharp absorption peak, short lifetime and efficient energy transfer, makes the J-aggregates ideal light sensitizers, e.g. in color photographic films. Hence there has been a strong and long-standing interest in the control of their photophysical properties. Both from quantum mechanical calculations as well as experimentally, it became clear that the excited state in an aggregate is not confined to a single molecule, but extends over many molecules (*18*). This so-called exciton is delocalized over several 10s of molecules (*19*), and thus the size of a J-aggregate determines the photophysical properties. Is the structural size of a single J-aggregate below the maximum exciton length, the absorption and fluorescence wavelength and bandwidth may depend on the size. A structural size above the exciton length should not influence the spectral properties.

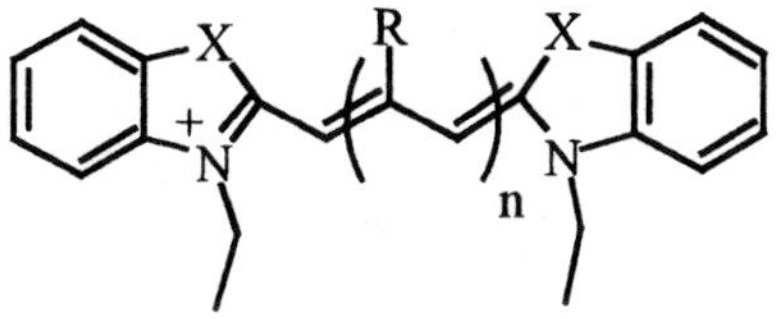

NK77: X=S, R=CH$_3$, n=1
NK85: X=O, R=H, n=1

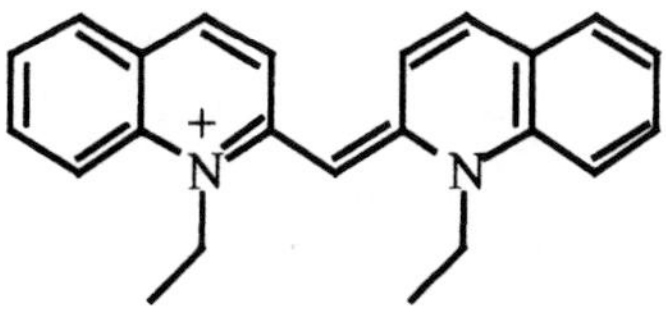

PIC

Figure 3. Chemical structures of cyanine dyes used in this study.

In the past there have been attempts to produce well-defined J-aggregates in thin polyvinylalcohol films. But Atomic Force Microsocopy (AFM) and Near-Field Optical Microscopy (SNOM) revealed that the nanometer-size primary J-aggregates agglomerate and form fiber-like super-aggregates with a length of several 100 μm (20). Even though the spectral properties can be determined, the fibrous structure does not allow the measurement of a single J-aggregate. By using dewetted polystyrene films with a micrometer-sized dome structure, on the other hand, it should be possible to produce J-aggregates with a more defined size, since the formation of the J-aggregates is limited to the volume of a single dome. Since the domes have submicrometer dimensions, the volume of a single dome is in the order of a few attoliter. Thus it contains a limited number of molecules. In order to realize the incorporation of J-aggregates into the polymer domes, aliquots of the dye solution in chloroform were mixed with the polymer solution to result in dye/polymer ratios of 0.1 to 15 wt%. Casting these mixed solutions onto substrates leads to the previously described dewetting and the formation of micron sized polymer domes.

Figure 4. Fluorescence micrograph of a dewetted sample containing 4 wt% of NK85 in polystyrene. The excitation wavelength is 440-480 nm. The inset is the Fast Fourier Transformation of the largest square of the fluorescence image.

(See third page in color insert.)

Fluorescence microscopy proved that the cyanine is incorporated into the polymer domes, since the area in between the domes does not fluoresce. The orange-red color of the fluorescence light in Figure 4 stemming from the domes is a first indication for the formation of J-aggregates. The domes have a small size distribution and are regularly spaced. The Fast Fourier Transformation image shows a halo with 6 spots which are an indication of a short range pseudo hexagonal order of the domes. Figure 5 shows the formation of J-aggregates of two other cyanine dyes.

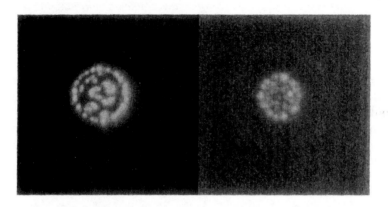

Figure 5. Fluorescence micrographs of polystyrene domes containing 4 wt% of NK85 (left) and PIC (right). The dome diameter is 6 μm in both cases. (See third page in color insert.)

It can be seen that each dome contains a countable number of J-aggregates. This implies that the number of interacting J-aggregates can easily be controlled by the concentration and the dome size. Lower concentration and smaller domes, respectively, lead to a smaller number of aggregates per dome. At very low concentrations, dewetted samples show another interesting effect of the dome size: even at the same concentration, smaller domes do not contain J-aggregates, whereas larger ones do (see Figure 6). From these observations it became clear that not only the concentration of the dye determines the aggregation state, but also that the number of interacting dyes in the microenvironment is a contributing factor to the aggregation. Thus it became possible to produce single J-aggregates which are isolated from each other in a well-defined environment. This effect can only be seen in non-homogeneous, microstructured samples, like the dewetted domes.

Figure 6. Fluorescence micrographs of a polystyrene dome containing 0.1% of NK77, which forms a single fluorescing aggregate. The domediameter is 12 μm.

(See fourth page in color insert.)

The concentration of the dye in the original casting solution is low, so that the dye exists in the molecularly dispersed, that is unaggregated state. But immediately after the solvent evaporation J-aggregates are present. This implies that during the casting process (which is a complex spatio-temporal process involving solvent evaporation, a fingering instability and "budding" of the polymer domes), the J-aggregates are formed. Hence the interplay of two different processes, far-from-equilibrium dewetting and towards-equilibrium aggregation, may also lead to interesting self-organized nano- and microstructures. Figure 7 shows examples where hierarchic structures can be formed. In some of the domes, the dye aggregates form ring structures in the vicinity of the edge of the polymer dome, an effect that also can be seen in Figure 5.

The origin of this ring formation is not fully understood, but the following mechanisms may reasonably explain the results. Dye aggregation is a type of crystallization and thus needs a nucleus to proceed. Nuclei form easily at interfaces and the three-phase-line (the contact line between the solid substrate, the liquid dome and air) is most likely to provide the defects at which nuclei are formed and aggregation proceeds. Another possibility is that nuclei are randomly formed across the entire volume of the dome. Then convection and material flow leads to an outward movement of the nuclei to the edge of the dome, similar to the "coffee stain" formation, which leads to deposition of ring-like deposition of coffee powder after a coffee spill. A third explanation is that the surface tension or capillary forces of the polymer dome drives the dye aggregates off the surface to the edge. Due to the easy formation of excitons after light irradiation, these spatially controlled arrangements of J-aggregates on the (sub) micrometer length scale may show interesting properties and may be useful for applications which need fast and efficient energy migration and electron transfer, like in solar cells, or for artificial photosynthesis.

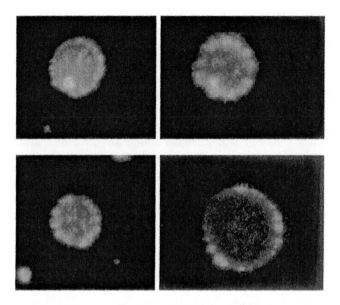

Figure 7. Fluorescence micrographs of domes where the self-organization of cyanine dyes occurred at the dome edge, leading to hierarchic structures. The diameter of the domes is in the range of 5-12 μm. (See fourth page in color insert.)

Dye Crystallization

The above mentioned J-aggregates are a rather specific type of molecular aggregates, and in order to extend the applicability of dewetted polymer patterns, the general phenomenon of crystallization of dyes in polymer matrices was investigated. Similar to the J-aggregates, dye crystals, often have interesting optic and photonic properties, for example the frequency doubling of light. This so called second harmonic generation (SHG) is a nonlinear optical (NLO) property which requires noncentrosymmetric arrangement of the NLO material (21). Molecularly dispersed dye molecules with a large dipole moment can be used as NLO material in thin films after poling in a strong electric field (typically 10,000 V/cm), but the films are metastable and the molecules tend to reorient to a NLO inactive random orientation. Another approach is the crystallization of the dye in non-centrosymmetric crystals. Defect-free large single crystals are difficult and time consuming to obtain.

Here we show that by co-casting of a dye solution and a polymer solution, or even by casting the dye solution only, micrometer-sized domes can be prepared and the crystallization of the dye within each individual dome can be controlled. Figure 8 summarizes the chemical formulae of the used NLO active dyes. Figure 9 is a fluorescence micrograph of a 50 %(wt/v) of **187** in polystyrene. Even though the dye makes up 50 % of the sample weight, the pattern formation is dominated by the fingering instability caused by the polystyrene in the solution. A control experiment of casting the dye solution only leads to the random crystallization during solvent evaporation and to the deposition of polycrystalline dye aggregates with a broad size distribution.

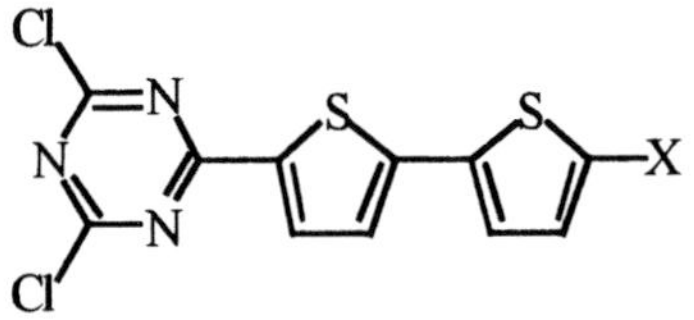

$$185: X = S\text{-}C_2H_5$$
$$187: X = N\text{-}(C_2H_5)_2$$

Figure 8. Chemical formulae of the investigated NLO dyes.

The close-up view of a single polymer dome shows that the dye is indeed crystalline. The dendritic shape of the crystal, which appears red in the fluorescence image, indicates the diffusion limited growth of the dye crystal after the polymer dome formed its spherical shape. Crystallization seems to have originated in the lower right part of the dome, which then spread through the dome to give a dendritic crystal. Obviously, the crystals cannot be larger than the dome itself, and thus the crystallization of dyes in mesoscopically restricted environments is interesting both, for the basic understanding of crystal growth as well as for the preparation of crystals with a well defined size and shape for photonic applications. Attempts to form mesoscopic domes of the dye **187** itself failed, since the dye crystallizes readily upon solvent evaporation. Thus, crystals are already formed in solution, and no pattern can be obtained. But other dyes of the same family can form metastable glasses upon rapid cooling, which was confirmed by differential scanning calorimetry (DSC). When those dyes are cast from solution, even though no polymer is present, a fingering instability leads to the formation of micron sized dome structures. Upon annealing the dye molecules in each dome crystallize. Depending on the annealing conditions (temperature, time) multicrystalline domes, single microcrystals, or fibrous crystals can be formed (*22*).

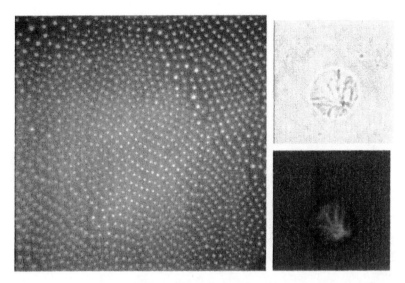

Figure 9. Micrographs of 187 in polystyrene (50 wt%). Monochrome fluorescence image of the pattern (left) and close up of a single dome (right), in transmission and fluorescence mode. The diameter of the domes are approx. 7 μm.
(See fifth page in color insert.)

Polarizing microscopy can be used to distinguish the aggregation state. Amorphous samples appear dark between crossed polarizers, polycrystals show various colors, which stems from the birefringence of the crystals and single crystals appear with a homogeneous color.

By investigating the temperature dependence of the crystal morphology, it was found that annealing at approx. 30 K below the bulk melting point of the compound, crystal fibers were formed. Annealing 50 K below bulk melting resulted in single crystals and annealing at 70 K below melting point gave polycrystalline domes (Figure 10).

Even though microphotolithography is a powerful tool to pattern surfaces and thin films with submicrometer precision, the spatiotemporal dynamics of dewetting and related self-organization processes have advantages. First, the patterning is governed by physical processes in an evaporating solution, like convection and fluctuations in surface tension. Thus the chemical nature of the solvent or the solute plays only a minor role in the patterning and a wide variety of compounds can be patterend. Second, the dynamics in the evaporating solution may lead to superstructures and hierarchic patterns in one step, something which is impossible by lithography alone. It also has to be mentioned that dewetting of a solution is a rapid and cost effective way of patterning and has the potential to be applied to large areas. Our present results show that dewetting induced pattern formation of polymers and low molar mass compounds now has reached a stage where these patterns may be applied to photonics devices.

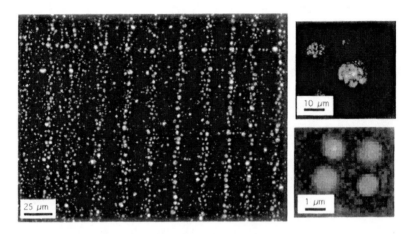

Figure 10. Polarizing micrograph of a dewetted sample of 185 on a glass substrate. Overview of the pattern (left). Polycrystalline dome prepared by annealing at room temperature (top right) and single crystalline domes prepared by annealing at 50°C (bottom right).

(See fifth page in color insert.)

References

1. Sharma, A. *Langmuir* **1993**, *9*, 861.
2. Reiter, G. *Langmuir* **1993**, *9*, 1344.
3. de Gennes, P. G. *Rev. Mod. Phys.* **1985**, *57*, 827.
4. Brochard-Wyart, F.; Daillant, J. *Can. J. Phys.* **1990**, *68*, 1084.
5. Brochard-Wyart, F.; de Gennes, P. G.; Hervert, H.; Redon, C. *Langmuir* **1994**, *10*, 1566.
6. Faldi, A.; Composto, R. J.; Winey, K. I. *Langmuir* **1995**, *11*, 4855.
7. Stange, T. G.; Evans, D. F.; Hendrickson, W. A. *Langmuir* **1997**, *13*, 4459.
8. Karthaus, O.; Grasjo, L.; Maruyama, N.; Shimomura, M. *Chaos*, **1999**, *9*, 308-314.
9. Fanton, X.; Cazabat, A. M.; Quare, D. *Langmuir* **1996**, *12*, 5875.

10. Scriven, L. E.; Sterling, C. V. *Nature* **1960**, *187*, 186.
11. Fanton, X.; Cazabat, A. M. *Langmuir* **1998**, *14*, 2554.
12. Shimomura, M.; Matsumoto, J.; Nakamura, F.; Ikeda, T.; Fukasawa, T.; Hasebe, K.; Sawadaishi, T.; Karthaus, O.; Ijiro, K. *Polym. J.* **1999**, *31*, 1115-1120.
13. Karthaus, O.; Koito, T.; Maruyama, N.; Shimomura, M. *Mol. Cryst., Liq. Cryst.* **1999**, *327*, 253-256.
14. Hellmann, J.; Hamano, H.; Karthaus, O.; Ijiro, K.; Shimomura, M.; Irie, M. *Jpn. J. Appl. Phys.* **1998**, *37*, L816-L819.
15. Jelly, E. E. *Nature* **1936**, *138*, 1009.
16. Scheibe, G. *Angew. Chem.* **1936**, *49*, 563.
17. *J-aggregates*; Kobayashi T. , Ed.; World Scientific; Singapore, 1996.
18. Davidov, A. S. *Theory of Molecular Excitons*; Plenum Press: New York, NY, 1971.
19. Minoshita, K.; Taiji, M.; Misawa, K.; Kobayashi, T. *Chem. Phys. Lett.* **1994**, *218*, 67.
20. Higgins, D. A.; Kerimo, J.; Bout, D. A.; Reid, P. J.; Barbara, P. F. *J. Am. Chem. Soc.* **1996**, *118*, 4049.
21. *Nonlinear Optical Properties of Organic Molecules and Crystals*; Charma, D.S.; Zyss, J. Eds.; Academic Press; New York, NY, 1987.
22. Karthaus, O.; Imai, T.; Miyakawa, D.; Watanabe, M. *Dewetting-Assisted Micro-Crystallization of Dyes*; in *Nanotechnology toward the Organic Photonics*; Sasabe, H. Ed.; GooTech; Chitose, Japan, 2002.

Chapter 17

Nanometer Expansion and Contraction Dynamics of Polymer Films Induced by Nanosecond Laser Excitation

H. Masuhara, T. Tada, and T. Masubuchi

Department of Applied Physics and Handai Frontier Research Center, Osaka University, Suita 565–0871, Japan

Intense excimer laser excitation of poly(methyl methacrylate) (PMMA) and polyimide films gives the transient expansion and the following contraction, which was directly measured as a function of nanosecond delay time during and after excitation. Depending on laser excitation condition, very rapid decay component and oscillatory behavior were observed in contraction and also phase transition is confirmed for PMMA. The behaviors are well interpreted by photothermal mechanism. In case of polyimide the expansion behavior is inconsistent with the simulation based on photothermal mechanism, on which we consider photochemical processes. New aspects of nonlinear polymer dynamics are presented and summarized.

Introduction

Studies on intense laser excitation of organic materials have received much attention as an interesting interdisciplinary research on morphological dynamics (*1-4*). In nanosecond excimer laser irradiation experiments, high density excited states are formed in the surface layer of the material (*5-9*), leading to efficient annihilation due to mutual interactions (*7, 8*), cyclic multiphoton absorption (*10*), and maybe multiphoton ionization (*9*) during the excitation pulse. As a result photothermal conversion takes place very rapidly in general (*6, 8, 10, 11*), while photochemical processes are involved for some polymers (*12, 13*). All the processes are in principle nonlinear behavior with respect to excitation photon number and lead to morphological changes such as expansion, melting, etching, and so on (*15-23*). The nature of these laser-induced dynamics can be elucidated by analyzing time-resolved observation of morphological changes in addition to time-resolved spectroscopic data providing the information on excited electronic states and photochemical processes. Indeed we have utilized fluorescence (*5-9*) and UV-visible absorption spectroscopy (*6, 24-28*), shadowgraphy (*30-33*), interferometry (*20-23*), surface light scattering imaging (*34*), polarization microscopy (*35*), photoaccoustic measurement (*27, 28*), and so on, all of which have nanosecond time resolution. In some cases we have succeeded in demonstrating directly how electronic excitation evolves to etching by combining spectroscopic and morphological data.

In the present review, we focus our attention to the laser-induced expansion and contraction dynamics of PMMA and polyimide and show their nonlinear property (*14, 29*). The expansion and contraction behavior of both films upon excimer laser irradiation has been measured as a function of the laser fluence by means of nanosecond interferometry. Below the ablation threshold, etching is not observed but the film undergoes rapid expansion during the excitation pulse and contracts through cooling. In case of high fluence, the contraction does not recover the original flat surface, which leads to permanent swelling. Expansion due to photochemical decomposition are occasionally involved, while phase transition and its dynamics were confirmed for the first time. The results on the morphological dynamics are obtained only by applying pulsed laser excitation and optical measurements, and will contribute to open new interdisciplinary research area of nonlinear dynamics of polymer systems.

Experiment

Material

PMMA (Kuraray Co. Ltd.) with a weight-averaged molecular weight of 102600 was used without further purification. Sample films were prepared by spin-coating a 15 wt % chlorobenzene solution of PMMA on quartz substrates. The film was baked for 2 h at 80°C to remove the residual solvent, giving a film thickness of ca. 2 μm. Its absorbance at 248 nm was 0.0015, and no appreciable absorption band of chlorobenzene was detected.

A N-methyl-2-pyrrolidone solution of the precursor polymer (Nissan Chemical Ltd.) was spin-coated onto quartz substrates, and heated for 60 min at 250 °C to induce polymerization. The chemical structure of the prepared polyimide is given in Scheme 1. Thickness of the film was ca. 1.5 μm, and absorption coefficient of the film was 37.2 μm^{-1} and 1.21 μm^{-1} at 248 nm and at 351 nm, respectively.

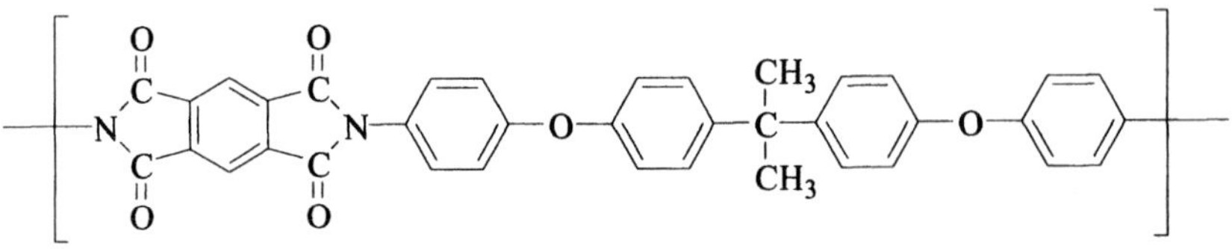

Scheme 1. Chemical structure of the used polyimide.

Laser Excitation

A KrF excimer laser (Lambda Physik LEXTRA 200, 248 nm, 30 ns fwhm) or a XeF excimer laser (ibid., 351 nm, 30 ns fwhm) was used as an excitation pulse for inducing expansion/contraction dynamics. The fluence was adjusted with partially transmitting laser mirrors, and was monitored shot-by-shot by a photodiode whose output was corrected with a joulemeter (Gentec, ED-200) with an oscilloscope (Hewlett-Packard, HP54522A). A central area of the excimer laser pattern with a homogeneous intensity distribution was chosen with an appropriate aperture and then focused onto the sample surface by using a quartz lens (f = 200 mm). Fresh surface of the sample film was used in every measurement. Etch depth was measured by a surface depth profiler (Sloan, Dektak 3). All experiments were done in air at room temperature.

Nanosecond Time-resolved Interferometry

The system of nanosecond time-resolved interferometry applied here is shown in Figure 1 (20-23). The second harmonic pulse of a Q-switched Nd^{3+}:YAG laser (Continuum Surelite I, 532 nm, 10 ns fwhm) was used as a probe light for the Michelson-type interferometer to measure excimer laser-induced morphological changes of the present film. Interference patterns were acquired by a CCD camera. Time-resolved measurement was carried out by controlling the delay time ($\triangle$t) between excitation and probe laser pulses with a digital delay/pulse generator (Stanford Research System, DG 535), and the delay time was monitored shot-by-shot by a digital oscilloscope. Here we define $\triangle$t = 0 as the time when the peaks of both laser pulses coincide with each other. All data were obtained by one-shot measurement to avoid effects by exciting photoproducts formed by previous irradiation.

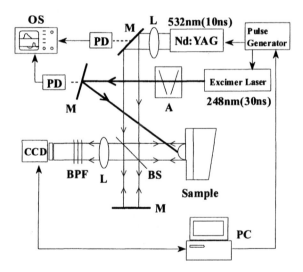

Figure 1. Time-resolved interferometry system. A 248 nm or 351nm excimer laser pulse (fwhm 30 ns) as as excitation light and a 532 nm pulse (fwhm 10 ns) as an probe light are irradiated to a sample film. Abbreviations are PD; photodiode, OS; oscilloscope, BS; beam splitter, PG; pulse generator, A; attenuater, L; lens, M; mirror, BPF; band pass filter.

Results and Discussion

Etch Depth

Upon excimer laser irradiation with high fluence, polymer films undergo permanent morphology changes; etching and swelling. Fluence dependence of the morphology change of PMMA film is given in Figure 2, where no change was observed below 800 mJ/cm^2, etching was left above 1400 mJ/cm^2, and permanent swelling was observed between these fluences. Therefore, the ablation and the swelling thresholds were determined to be about 1400 mJ/cm^2 and 800 mJ/cm^2, respectively.

Fluence dependence of etch depth of polyimide film is also given in Figure 2, where the ablation threshold at 248 nm excitation was determined to be about 40 mJ/cm^2 and the ablation and the swelling threshold at 351 nm excitation were about 210 mJ/cm^2 and 150 mJ/cm^2, respectively. At 248 nm excitation the etch depth increases monotonously with the laser fluence. On the other hand, the film shows a permanent swelling from about 150 mJ/cm^2 to 210 mJ/cm^2 at 351 nm excitation, and undergoes laser ablation above the latter fluence. Enormous difference in the ablation threshold, depending on the excitation wavelength, may be ascribed to absorption coefficient at the excitation wavelength. It is also expected that ablation mechanism due to photochemical and photothermal degradation of the polymer depends on energy of excitation photons.

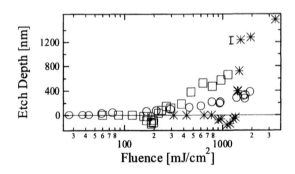

Figure 2. Etch depth profile of PMMA film obtained with 248 nm (*) excimer laser irradiation and that of polyimide film obtained with 248 nm (○) and 351 nm (□) excimer laser irradiation. Negative etch depth means permanent swelling. An error bar is included.

Expansion and Contraction Dynamics of PMMA Film

In Figure 3, transient expansion and contraction dynamics of PMMA film was given at the fluence of 700 and 540 mJ/cm^2. The irradiated film began to expand at the late stage of the excimer laser pulse, and the expansion disappeared completely via slow contraction whose time constant was about 50 μs. Namely, the original flat surface was perfectly recovered, and we cannot discriminate the original and recovered surfaces by our interferometry. It is worth noting that transient expansion and rapid contraction gave bump around ~100 ns at the fluence of 700 mJ/cm^2, while it was not observed at the fluence of 540 mJ/cm^2. Such a behavior was never observed in transient expansion and contraction dynamics of PMMA film doped with aromatic molecule (*20, 21*).

Similar interferometric measurements were performed at the fluence of 400, 300, and 250 mJ/cm^2 and summarized in Figure 4. At the fluence of 300 and 250 mJ/cm^2, the film surface showed repetitive expansion and contraction of several tens nm, reached a plateau value, and then recovered to the original flat surface. At the fluence of 400 mJ/cm^2, the expansion amplitude and the oscillatory period of the film seemed to increase. It is worth noting that oscillatory behavior of expansion and contraction depends on the laser fluence and reflects nonlinear polymer response.

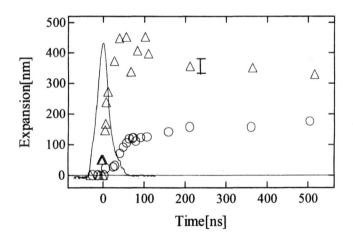

Figure 3. Expansion and contraction dynamics of PMMA film at the fluence of 700 mJ/cm^2 (△) and 540 mJ/cm^2 (○) below the swelling threshold. A solid curve represents an excimer laser pulse, and an error bar is included.

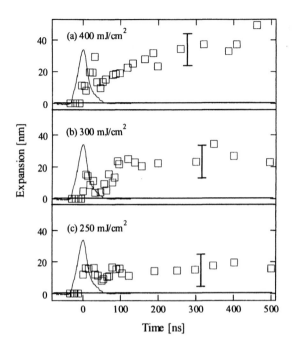

Figure 4. Expansion and contraction dynamics of PMMA film at the fluence of (a) 400 mJ/cm², (b) 300 mJ/cm², and (c) 250 mJ/cm² below the fluence of 450 mJ/cm² at which glass-rubber transition of PMMA is brought about. A solid curve represents an excimer laser pulse, and an error bar is included.

Possible Expansion Mechanisms of PMMA Film

It is well known that PMMA decomposes into monomer, CO_2, and so on upon excitation, so that these gaseous molecules may cause expansion. Concerning ablation of PMMA, Garrison and Srinivasan reported that gaseous monomer products from PMMA result in a volume increase of 30% under a certain condition (36), which seems consistent with the present result. SEM observations of PMMA films containing pyrene and cholobenzene (37) and of those containing IR-165 and diazo Meldrum's acid (13), which are irradiated by multi laser shots show that various holes are left. This result was interpreted by assuming that formed gas molecules leak to air, leaving holes. In the present intereferometric images, however, flat surface was always maintained during expansion at the fluence below the threshold of permanent swelling. If a lot of holes whose size are larger than wavelength of probe light are formed by ejection of gaseous molecules upon contraction, clear and bright images should not be observed. Whether the film undergoes rupture or not depends on experimental conditions such as excitation wavelength, irradiation pulse number, pulse width, and existence of dopant. Thus, we consider that polymer decomposition mechanism leading to volume change may not have a major role in the present expansion and contraction.

Another possible mechanism is ascribe to solvent chlorobenzene upon excitation. If appreciable amount of solvent molecules are left in the film, they evaporate and result in volume increase of the film. Also photodissociation of chlorobenzene to Cl atom and counter radicals may involve formation of gaseous molecules. However, as mentioned above, it is confirmed directly by UV absorption spectral measurements that chlorobenzene absorption is negligible compared to PMMA. Hence, solvent role in expansion should be excluded, and we consider photothermal processes are responsible to the present expansion phenomena of PMMA.

Photothermal Mechanism and Glass-Rubber Transition of PMMA Film

To examine the nature of photothermal expansion and contraction behavior of PMMA film, it is important to correlate the transient dynamics to thermal properties. We measured fluence dependence of the expansion of PMMA film at the delay time of 1 μs after the oscillatory behavior is finished. It is shown in Figure 5 that the expansion of PMMA film consists of two components having different slopes. Around the fluence of 450 mJ/cm^2, a knick point was observed and the slope increased by a factor of about 8.6. This suggests that physical and chemical nature attained at the delay time of 1μs is different between two regions. We consider that the area irradiated by the excimer laser was rapidly heated through efficient photothermal conversion processes. This is well consistent with photothermal temperature elevation of doped PMMA film where nanosecond heating during the pulse was directly confirmed by fluorescence spectral broadening in a few tens ns (*6, 8*). The temperature elevation should result in changes of physical properties of polymer films, and the temperature of PMMA film is estimated at the critical fluence of 450 mJ/cm^2. Assuming that multiphoton absorption and saturation in absorption are excluded and absorbed energy is converted to heat with unit quantum yield, and considering errors in small absorbance at 248 nm, it is concluded that the temperature attained at 450 mJ/cm^2 is roughly the same to Tg of PMMA (*38*) (378 K). Therefore, it is considered that the different expansion amplitudes states of PMMA film, respectively.

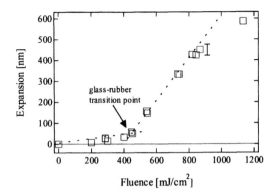

Figure 5. Fluence dependence of the expansion at 1 µs after the excimer laser irradiation of PMMA film. An error bar is included.

If effective absorbance and photothermal conversion efficiency are independent on the fluence, absorbed energy should be proportional to the latter and the slopes can be discussed simply in terms of thermal expansion coefficient of polymer films. The coefficient of linear expansion of rubber and glass states, α_g and α_r, below and above glass-rubber transition temperature (Tg) of PMMA is $(2.5\text{-}2.7) \times 10^{-4}$ K^{-1} and $(5.6\text{-}5.8) \times 10^{-4}$ K^{-1}, respectively (*38*). We consider that the one-dimensional expansion is induced along the perpendicular direction to the film, since polymer film outside the area irradiated by the excimer laser is of course hard. Then, the volume expansion could be replaced by the linear one along the thickness, so that $(\alpha_r/\alpha_g)^3$ corresponds to the change in the slope below or above the knick point. Indeed, $(\alpha_r/\alpha_g)^3$ is in agreement with the slope ratio. Not only thermal expansion coefficient but also specific heat, elastic modulus, and so on, are much influenced by the glass-rubber transition (*39, 40*). This may raise a difficulty in quantitative analysis, while the present explanation seems quite reasonable.

The present result reminds us that the different expansion behavior between doped PMMA and poly(methyl acrylate)(PMA) films is explained in relation to glass-rubber transition (*21*). The former and latter films are in glass and rubber states at room temperature, and gave smaller and larger expansion coefficients, respectively. It was also proved that the expansion of pyrene-doped PMMA film in the glass state is faster compared to that involving glass-rubber transition at the higher fluence (*21*). The results were interpreted in terms of phase transition dynamics and efficient photothermal conversion processes of pyrene under high excitation condition. Indeed it was demonstrated directly that the phase transition from glass to rubber takes place in 10 ~ 20 ns. The present result on the dynamics of neat PMMA film can be explained also as photothermal expansion and contraction, and the phase transition, which shows a new aspect of nonlinear morphological properties of polymer dynamics.

Expansion and Contraction Dynamics of Polyimide Film

At 248 nm excitation, expansion and contraction dynamics of polyimide film is given in Figure 6 at the fluence of 20 mJ/cm^2. The irradiated film began to expand during the excimer laser pulse and contracted rapidly, whose rate was about 1 nm/ns and 0.1 nm/ns, respectively. The maximum expansion amplitude was attained when the excitation pulse ends, and the expansion disappeared completely after a few ms.

Expansion dynamics at 351 nm excitation with the fluence of 80 mJ/cm^2 and 130 mJ/cm^2 below the ablation threshold is shown in Figure 7. The irradiated film began to expand during the excimer laser pulse similarly as at 248 nm excitation, and expansion amplitude increased with the fluence. The contraction took place quite slowly, except small bump at 130 mJ/cm^2, and then the original flat surface was recovered.

First we consider that the different expansion behavior at 248 nm and 351 nm excitation is due to the different etching mechanism discussed above. In the case of 351 nm excitation, multiphoton photochemical and photothermal processes are involved and the latter may be more important at lower fluence. The contraction behavior at 80 mJ/cm^2 may reflect slow heat dissipation in polyimide, which behavior is actually consistent with photothermal expansion and contraction dynamics of PMMA film (21). On the other hand, the thin surface layer is excited at 248 nm, so that heat dissipation to quartz substrate should be slower than that at 351 nm. More quantitative analysis of cooling processes by the simulation was conducted to understand the expansion and contraction dynamics.

Simulation of Surface Temperature Elevation of Polyimide Film

Here we assume photothermal mechanism is responsible even for 248 nm excitation, and estimate numerically temperature rise and decay dynamics. As large excitation energy is confined at the thin surface layers of 27 nm and 830 nm thickness for 248 nm and 351 nm excitation, respectively, assuming Lambert-Beer law hold, temperature elevation is extremely large. Here the surface of the 20 nm thickness is numerically examined for both excitation wavelengths. Surface temperature rise as a function of delay time was calculated by solving one-dimensional heat conduction equation. In calculation it was

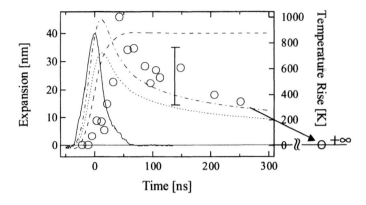

Figure 6. Expansion and contraction dynamics of polyimide film at the fluence of 20 mJ/cm² below the ablation. The solid and dashed curves represent the time profiles of the excimer laser pulse and time-integration of the laser pulse, respectively, while the dotted and dash-dotted curves are simulated surface temperature rise at the fluence of 20 mJ/cm² and 30 mJ/cm², respectively. Excitation wavelength is 248 nm. An error bar is included.

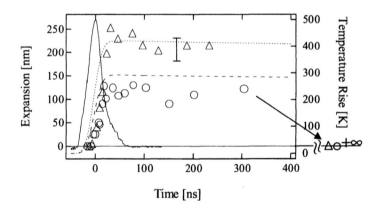

Figure 7. Expansion and contraction dynamics of polyimide film at the fluence of 80 mJ/cm² (○) and 130 mJ/cm² (△) below the ablation threshold. The solid curve represents the time profile of the excimer laser pulse. The dashed and dotted curves represent the simulated surface temperature rise at the fluence of 80 mJ/cm² and 130 mJ/cm², respectively. Excitation wavelength is 351 nm. An error bar is included.

assumed again that absorption coefficient of α, heat conductance, and density are independent upon both temperature and excitation intensity, Lambert-Beer law holds, and absorbed photon energy is fully converted to heat. We used heat conductance of 1.20×10^{-3} W/cm· K(41), density of 1.42 g/cm^3 (42) for polyimide, and the following temperature dependent specific heat (43):

$$C_P(T) = 0.96 + 1.39((T - 300)/400) - 0.43((T - 300)/400)^2$$

where T is temperature.

The simulated surface temperature change for each excitation condition is given in Figures 6 and 7. Upon 248 nm excitation, the surface temperature reaches 1010 K and 1280 K at the fluence of 20 mJ/cm^2 and 30 mJ/cm^2, respectively. On the other hand the degradation temperature of polyimide is experimentally determined to be about 700 K (44). It is noticeable that the calculated temperature is higher than 700 K but attained only for a few hundreds ns. This suggests that photothermal degradation is not probable, as polyimide degradation at 700 K was confirmed by 100 min heating. In the case of 351 nm excitation, the estimated peak temperature is 690 ~ 860 K at the fluence of 80 ~ 130 mJ/cm^2. As temperature decreases by 100 K takes only 400 ns, thermal degradation again seems impossible also in this case. The expansion and contraction dynamics is qualitatively consistent with the simulated temperature change, particularly at 80 mJ/cm^2. This is quite reasonable as expansion amplitude should be proportional to temperature elevation. Thus morphological dynamics obtained at 351 nm excitation can be interpreted in terms of photothermal processes described above.

On the other hand, the rapid decay component observed at 248 nm excitation, was not reproduced by this simulation as is well demonstrated in Figure 6. Namely, photochemical decomposition may be induced and result in gaseous small molecules, and their leakage to air or recombination are probably responsible to the behavior. This explanation is consistent with the behavior just below the threshold.

Conclusion

Intense excimer laser irradiation of polymer films injects large amount of thermal energy or induces decomposition appreciably in a few tens ns, which triggers vigorous motion of polymers and unlace their interpenetration. As the motion of polymers is coupled with each other, the resultant morphological changes can be detected as nanometer expansion dynamics. This was achieved by following surface displacement induced by pulsed excitation with our nanosecond interferometry. While the contraction to the original surface place is

224

observed in microsecond-millisecond time range, some interesting expansion behaviors were demonstrated. Depending on laser wavelength, fluence, and polymer, expansion involving phase transition, oscillatory expansion and contraction, expansion with the bump, and so on were observed and their mechanisms were considered in terms of photothermal and photochemical viewpoints. New aspects on nonlinear dynamics of polymers are presented and summarized, which will contribute to open science and technology of soft materials.

References

1. Glezer, E. N.; Milosavljevic, M.; Huang, L.; Finlay, R. J.; Her, T.- H.; Callan, J. P.; Mazur, E. *Opt. Lett.* **1996**, *21*, 2023
2. Glezer,E. N.; Mazur, E. *Appl. Phys. Lett* **1997**, *71*, 882.
3. Tsuboi, Y.; Akita, H.; Yamada, K.; Itaya, A. *Jpn. J. Appl. Phys.* **1997**, *36*, L1048.
4. Gery, G.; Fukumura, H.; Masuhara, H. *J. Phys. Chem. B* **1997**, *101*, 3698.
5. Tsuboi, Y.; Fukumura, H.; Masuhara, H. *J. Phys. Chem.* **1995**, *99*, 10305.
6. Fujiwara, H.; Nakajima, Y.; Fukumura, H.; Masuhara, H. *J. Phys. Chem.* **1995**, *99*, 11481.
7. Fukumura, H.; Takahashi, E.; Masuhara, H. *J. Phys. Chem.* **1995**, *99*, 750.
8. Fujiwara, H.; Fukumura, H.; Masuhara, H. *J. Phys. Chem.* **1995**, *99*, 11844.
9. Tsuboi, Y.; Hatanaka, K.; Fukumura, H.; Masuhara, H. *J. Phys. Chem.* **1994**, *98*, 11237.
10. Fukumura, H.; Masuhara, H. *Chem. Phys. Lett.* **1994**, *221*, 373.
11. Fukumura, H.; Mibuka, N.; Eura, S.; Masuhara, H.; Nishi, N. *J. Phys. Chem.* **1993**, *97*, 13761.
12. Küper, S.; Brannon, J.; Brannon, K. *Appl. Phys. A* **1993**, *56*, 43.
13. Hahn, Ch.; Lippert, T.; Wokaun, A. *J. Phys. Chem. B* **1999**, *103*, 1287.
14. Masubuchi, T.; Furutani, H.; Fukumura, H.; Masuhara, H. *J. Phys. Chem. B* **2001**, *105*, 2518.
15. Kawamura, Y.; Toyoda, K.; Namba, S. *Appl. Phys. Lett.* **1982**, *40*, 374.
16. Srinivasan, R.; Mayne-Banton, V. *Appl. Phys, Lett.* **1982**, *41*, 576.
17. Srinivasan, R.; Leigh, W. J. *J. Am. Chem. Soc.* **1982**, *104*, 6784.
18. Srinivasan, R.; Baren, B. *Chem. Rev.* **1989**, *89*, 1303.
19. Miller, J. C. *Laser ablation- Principles and Applications*, Springer-Verlag: Berlin, 1994.
20. Furutani, H.; Fukumura, H.; Masuhara, H. *Appl. Phys. Lett.* **1994**, *65*, 3413.
21. Furutani, H.; Fukumura, H.; Masuhara, H. *J. Phys. Chem.* **1996**, *100*, 6871.
22. Furutani, H.; Fukumura, H.; Masuhara, H.; Lippert, T.; Yabe, A. *J. Phys. Chem. A* **1997**, *101*, 5742.

225

23. Furutani, H.; Fukumura, H.; Masuhara, H.; Kambara, S.; Kitaguchi, T.; Tsukada, H.; Ozawa, T. *J. Phys. Chem. B* **1998**, *102*, 3395.
24. Wen, X.; Tolbert, W. A.; Dlott, D. D. *Chem. Phys. Lett.* **1992**, *192*, 315.
25. Lee, I.-Y. S.; Wen, X.; Tolbert, W. A.; Dlott, D. D.; Doxtader, M. D.; Arnold, R.; *J. Appl. Phys.* **1992**, *72*, 2440.
26. Wen, X.; Tolbert, W. A.; Dlott, D. D. *Chem. Phys. Lett.* **1993**, *99*, 4140.
27. Hatanaka, K.; Kawao, M.; Tsuboi, Y.; Fukumura, H.; Masuhara, H. *J. Appl. Phys.* **1997**, *82*, 5799.
28. Tsuboi, Y.; Hatanaka, K.; Fukumura, H.; Masuhara, H. *J. Phys. Chem. A* **1998**, *102*, 1661.
29. Masubuchi, T.; Takuji, T.; Nomura, E.; Hatanaka, K.; Fukumura, H.; Masuhara, H. *J. Phys. Chem. A* **2002**, *106*, 2180.
30. Srinivasan, R.; Braren, B.; Cassey, K. G.; Yeh, M. *Appl. Phys.* **1989**, *55*, 2790.
31. Srinivasan, R.; Braren, B.; Cassey, K. G.; Yeh, M. *J. Appl. Phys.* **1990**, *67*, 1604.
32. Tsuboi, Y.; Hatanaka, K.; Masuhara, H. *Appl. Phys. Lett.* **1994**, *64*, 2745.
33. Bennett, L. S.; Lippert, T.; Furutani, H.; Fukumura, H.; Masuhara, H. *Appl. Phys. A* **1996**, *63*, 327.
34. Hatanaka, K.; Itoh, T.; Asahi, T.; Ichinose, N.; Kawanishi, S.; Sasuga, T.; Fukumura H.; Masuhara, H. *Appl. Phys. Lett.* **1998**, *73*, 3498.
35. Hosokawa, Y.; Mito, T.; Tada, T.; Tanaka, T.; Asahi, T.; Masuhara, H. *Proceeding of SPIE* **2002**, *4426*, 113.
36. Garrison, B. J.; Srinivasan, R. *Appl. Phys. Lett.* **1984**, *44*, 849.
37. Lippert, T.; Webb, R. L.; Langford, S. C.; Dickinson, J. T. *J. Appl. Phys.* **1999**, *85*, 1838.
38. Brandrup, J.; Immergut, E. H. *Polymer Handbook*, 3rd. ed.; John Wiley & Sons, New York, 1989.
39. Schwarz, G. *Cryogenics* **1988**, *28*, 248.
40. Sperling, L. H.; *Introcuction to Physical Polymer Science* ; Wiley & Sons: New York, 1986, Chapter 6.
41. Feurer, T.; Wahl, S.; Langhoff, H. *J. Appl. Phys.* **1993**, *74*, 3523.
42. Brunco, D. P.; Thompson, M. O.; Otis, C. E.; Goodwin, P. M. *J. Appl. Phys.* **1992**, *72*, 4344.
43. Kuper, S.; Brannon, J.; Brannon, K. *Appl. Phys. A* **1993**, *56*, 43.
44. Numata S.; Kinjo N. *Kobunshi Ronbunshu* **1985**, *42*, 7.

Chapter 18

Nonlinear Dynamics in Surfactant Systems

Mark Buchanan

Department of Complex Systems, Vrije Universiteit, Amsterdam,
The Netherlands

Interface instabilities, known as myelins, are an example of
exotic nonequilibrium behavior present during dissolution in a
number of surfactant systems. Although much is known about
equilibrium phase behavior much still remains to be
understood about nonequilibrium processes present in
surfactant dissolution. In this chapter nucleation and growth,
self and collective diffusion processes and nonlinear dynamics
and instabilities observed in various polymeric systems are
reviewed. These processes play an important role in our
understanding of myelin instabilities. Kinetic maps and the
concept of the free energy landscape provide a useful approach
to rationalize some of the more complex behavior sometimes
observed.

Although much work has been done to study equilibrium behavior in surfactant systems many nonequilibrium dynamic behavior are still far from well understood. When neat or concentrated surfactant is contacted with solvent complicated diffusion process occurs due to the presence of mesophase at the interface. Initially, at the interface, the formation, type and structure of the mesophase will influence the subsequent dynamics. In some cases the interface can become unstable during dissolution and rather striking instabilities form. To obtain a good understanding of such complicated nonlinear processes has relied on a systematic study of the equilibrium phase behavior in such systems. This has given us a firm basis on which to study the nonequilibrium behavior.

In this chapter I will review problems related to dissolution kinetics and nonlinear process that occur in surfactant systems. First we discuss the role of phase kinetics in surfactant dissolution. Then the diffusive processes are discussed where it is important to appreciate the difference between self and collective diffusion. Finally, interface instabilities will be discussed which includes some of the most recent and significant observations. These studies are extremely interesting in the context of industrial problems such as detergency.

Kinetics of phase formation

The dissolution of a mesophase is, in most cases, *diffusions limited*. This means the composition at the interface between phases corresponds to the equilibrium composition. This is indeed the main assumption behind the linear penetration scan which involves observing surfactant contacted with an aqueous phase in a capillary tube or between a glass slide and coverslip (2-9). Using polarization microscopy the structural arrangement of the intermediate mesophases can be identified (10). A quantitative approach to verify that the phase boundaries coincide with the equilibrium phase boundaries can be performed using refractive index measurements (11). It is true provided concentration gradients are not too large and the evolution is linear. Interferometry can also be used to follow the whole concentration profile as it evolves through time (12).

Generally mesophases form rapidly which leads to a diffusion-limited growth. Times of order seconds or less have been reported in T-jump experiments where a homogeneous sample is subjected to a temperature change and the time for the mesophase to form is measured (13). However, in some penetration scan experiments times much longer than a second have been observed.

Time resolved X-ray and neutron scattering have also been used to elucidate the kinetic behavior in many surfactant systems (14, 15). For the case of dissolution kinetics transitions between micelle and vesicle structures have been studied during homogeneous dilution of the solvent (16). Transient structures such as disks have been observed during such a transition.

The molecular assembly of a typical mesophase involves molecules having to diffuse distances of about $\lambda \sim 10$nm to make new structures. Given the self diffusion coefficient is of order $D \sim 10^{-11}$ m^{2}s^{-1} then we would expect structures to form $\lambda^2 / D \sim \mu$s – ms where as timescales of order seconds are observed. The accepted and most likely explanation is the need to nucleate the new phase.

The delay of one phase out of another can be explained by nucleation and growth in simple systems. When there is competition to nucleate several mesophases then it becomes too complicated to think in these simple terms. So for more complicated systems it is easier if we think in terms of the *free energy landscape*.

At this point it is useful mention work on colloid / non-adsorbing polymer mixtures (*17–19*). This simple system provides us with a case study where the nonequilibrium behavior can be successfully explained by kinetic maps determined from the *free energy landscape*.

Such a theory has also been useful is describing behavior of samples in the three-phase region (lamellar, sponge and micellar phases) (*20*). After applying a temperature change the phase boundaries change position. As the sample relaxes, most of the sponge phase is replaced by lamellar phase and the sample is observed as it equilibrates. All directly observable behavior can be explained in terms of a quasi-equilibrium free energy landscape by first identifying the fast and slow components. In this case, bilayer concentration equilibrates rapidly whereas the bilayers reorganize slowly.

Diffusion processes

During surfactant dissolution the two diffusion processes can be identified. On the molecular scale a molecule undergoes *self diffusion* where the diffusion coefficient is determined from its mean squared displacement. Various NMR techniques have been used to study quantitatively self diffusion processes (*21-24*). It is important to note that each component in the system will have a *self diffusion* coefficient. Diffusion coefficients for a number of mesophase systems have been collected where values of order $10^{-12} - 10^{-11}$ m^{2}s^{-1} were reported (*25*). The *self diffusion* coefficients of the solvent are typically reduced no more than an order of magnitude in the presence of mesophases which essentially act as obstacles to the solvent (*25, 26*).

Collective diffusion is the response of a given species to a concentration gradient. In the dissolution process collective diffusion plays the most important role. In the case of a two-component system there will be only one collective diffusion coefficient as a surfactant concentration gradient inevitably implies a solvent concentration gradient the opposite way.

Dynamic light scattering (DLS) is an effective technique to measure the collective diffusion coefficients by measuring the time correlation of the concentration fluctuations. These experiments allow us to can determine the collective modes of the system that couple to concentration fluctuations. In binary systems this connection is quite useful since there is only one independent concentration variable. Then one can obtain the collective diffusion coefficient theoretically from the dynamic structure factor in the long wavelength limit.

It is not always the case that self diffusion and collective diffusion coefficients are related. In the case of water transport in a lamellar phase the collective modes of motion, including the modes that correspond to concentration fluctuations have been identified (27). Since the water movement between bilayers can be considered as Poiseuille flow where the driving force is a pressure gradient in the water which is related to the mean interlayer spacing and hence to the surfactant concentration. For the case of a dilute lamellar phase this connection can be used to calculate the collective diffusion coefficient. This can be related to microscopic quantities such as solvent viscosity and the bilayer interaction energy such as electrostatics, van der Waals (27, 28), or undulation forces (29). Furthermore, in nonionic surfactant hexagonal phases collective diffusion coefficients have been found to be more than an order of magnitude larger than the self diffusion coefficient (30). In all of the above cases the samples are considered well oriented but in reality the phase would be constructed from randomly oriented domains which complicates further this senario.

Dissolution kinetics

In most penetration scans performed in surfactant dissolution experiments the phases are homogeneous and the interface between them is sharp. However, in some cases the interface becomes unstable and dramatic instabilities can be observed. There are many examples of instabilities that are well understood that maybe rationalized in terms of kinetic maps or dissolution paths, or dynamic instabilities involving fluid flow (e.g. Marangoni effects) or other "Laplacian growth instabilities", such as Mullins-Sekerka instabilities (31). However, *myelins* (Figure 1) are an example of an instability that remains poorly understood.

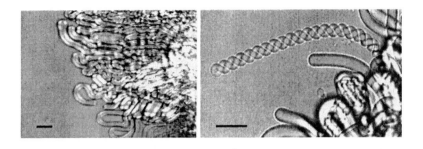

Figure 1. (a) Myelin interface in C12E3/water system; (b) At later times (several minutes) coil structures can be observed. Scale bar is 20μm.

Myelin formation is perhaps one of the most intriguing instabilities in nature. These instabilities appear during the swelling and dissolution of a surfactant lamellar phase and are primarily observed in surfactant systems that have a large miscibility gap between the lamellar phase and the solvent. In this case surfactant molecules can remain organized in bilayer structures up to high dilutions due to their low preferred curvature. (In contrast myelins are not observed in surfactant systems that have higher preferred curvature which posses a very small miscibility gap.) Within myelin the surfactants are organized in bilayer structures where the bilayers stack in a multi-tubular fashion. These multi-lamellar tubules and are approximately ten microns in width (Figure 1).

The myelin phenomenon started to appear in old articles published on solubility and detergency of soap solutions. Stevenson reported myelinic like structures during the removal of oil from clothes fibers (32). A few years later Harker also referred to myelin instabilities in similar experiments (33). In the late 1950s Lawrence reported qualitatively that myelins with different structures were dependent on the surfactant and composition (34, 35).

The first quantitative results were not produced until the 1980s when Sakurai and coworkers obtaining growth rates for the myelins in the egg-yolk lecithin/water system (36, 37). Using freeze fracture electron microscopy they have confirmed that the bilayers in the myelins have a multi-tubular organization (38, 39). They have also reported that myelins can have a variety of structures depending on the length of time they have been growing (40). The lamellar formation and structure and the influence of counterions have been studied using both microscopy and freeze fracture techniques (41, 42).

More advanced techniques have been used to study dynamics of myelins in penetration scans by using tracer particles to follow fluid flow (9). As well as quantification of the swelling process by determination of an effective diffusion coefficient many qualitative features have been observed.

Penetration scans where swelling and growth dynamics is observed up to longer timescales show a change in the growth exponent. At later times there is a transition from a diffusive to subdiffusive regime where the growth exponent changes from $t^{1/2}$ to about $t^{1/3}$ (Figure 2). Such a transition is likely to be due to some complicated internal reorganization of the bilayers (*43*).

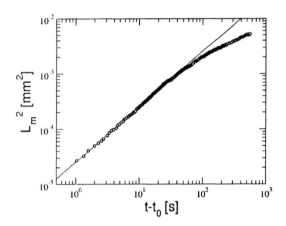

Figure 2. Temporal evolution of myelin growth in lecithin (Egg PC) system over long times reveals a diffusive regime at early times followed by a subdiffusive regime. Figure courtesy of J. Leng.

Another intriguing feature of myelin growth is there ability to form helical structures (Figure 1b). These have been observed in a variety of systems (*9, 40, 44, 45*). One argument for there formation is based on the influence of spontaneous curvature of the membranes in the myelin figures (*45*). By spraying polymer onto the surface of the myelin figure the spontaneous curvature can be changed and coiling is observed. A recent theoretical study has also shown that spontaneous curvature and a decrease in the bilayer spacing can also lead to coiling instability (*46*).

Myelins have also been observed to collapse when an intermediate sponge phase is present between the lamellar phase and the micellar phase (*47*). After the formation of the myelins sponge phase is observed to from at the surface (Figure 3). The delayed formation of the sponge phase in this case is consistent with the kinetic-pathway theory; initially the swelling of bilayers occurs and at later timescales bilayers can reorganize forming sponge phase. In this case is the fast and slow components can be identified to the bilayer concentration and reorganization respectively.

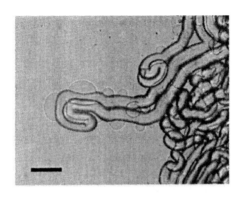

Figure 3. Deflated myelin in $C_{12}E_5$/ W penetration scan at T=60 °C where an intermediate sponge phase is present. After some time patches of sponge phase form at the myelin surface. Bar is 20μm.

Despite most of the myelin studies having focused on growth and late timescale behavior the mechanism for their formation is still unknown. Penetration scans using *onion phase* (lamellar phase is presheared into multilamellar vesicles or *onions*) have shown that myelin formation can be suppressed (*48*). This implies that there formation is sensitive to bilayer organization in the lamellar phase. Dissolution of these onion phases also have interesting and exotic behavior (*48, 49*).

Instabilities that manifest themselves in surfactant and polymeric systems have been considered in an attempted to elucidate the myelin instability. In polymer-like micelles (or wormlike micelles) instabilities have been observed in the directional growth of hexagonal phases in a temperature gradient (*30*). These instabilities are an example of the Mullins and Sekerka type (*31*). In the case of polymer gels, instabilities appear during growth, which resemble a raspberry like texture at the surface. This instability is due to the elastic properties of the gel which is a network of chemically bonded polymers. As the gel swells at the surface it remains anchored to the rest of the unswollen gel and the surface buckles (*50*).

However, myelin instabilities have evaded elucidation by the above mensioned instabilities. The well known finger-like instabilities which involves material diffusing onto the tip do not apply in this case as myelin growth as growth occurs as a result of backflow. Furthermore, myelins cannot be described by elastic instabilities such as polymer gels since there is no strong anchoring between layers in the lamellar phase. Myelins still remain unclassified in terms of our current understanding of instabilities in a variety of systems.

Conclusions

Our progress in understanding dissolution kinetics has been advanced from surfactant systems whose equilibrium phase behavior is well understood. Since phase formation is relatively fast, the kinetics are often simple and the dynamics are often controlled by collective diffusion. In cases where this breaks down and more complex behavior is observed free energy landscapes have provided us with important new insights. However, formation of instabilities such as myelins still remain a mystery and cannot be classified with well founded surfactant and polymeric instability theories. At present no theoretical explanation for their formation exists despite having been observed 150 years earlier by R. Virchow (*51*).

Acknowledgement

I would like to thank P. Warren, J. Leng, M. E. Cates, S. U. Egelhaaf and P. R. Garrett for useful discussions.

References

1. Laughlin, R. G. *The Aqueous Phase Behaviour of Surfactants*; Academic Press, 1994.
2. Warren, P. B.; Buchanan, M. *Curr. Opin. Colloid Interface Sci.* **2001**, 6, 287-293.
3. Miller, C. A. *Colloids Surf., A* **1988**, 29, 89–102.
4. Miller, C. A.; Raney, K. H. *AIChE J.* **1987**, 33, 1791–1799.
5. Miller, C. A. *Tenside, Surfactants, Deterg.* **1996**, 33, 191.
6. Miller, C. A.; Raney, K. H. *Colloids Surf., A* **1993**, 74, 169–215.

7. Mori F.; Lim, J. C.; Raney, O. G.; Elsik, C. M.; Miller, C. A. *Colloids Surf., A* **1989**, 40, 323–345.
8. Hakemi, H.; Varanasi, P. P.; Tcheurekdjian, N. *J. Phys. Chem.* **1987**, 91, 120–125.
9. Buchanan, M.; Egelhaaf, S. U.; Cates, M. E. *Langmuir* **2000**, 16, 3718–3726.
10. Kleman, M. *Points, Lines and Walls.* Wiley-Interscience Publication, 1983.
11. Laughlin, R. G. *Adv. Colloid Interface Sci.* **1992**, 41, 57–79.
12. Chen, B-H.; Miller, C. A.; Walsh, J. M.; Warren, P. B.; Ruddock, J. N.; Garrett, P. R.; Argoul, F.; Leger, C. *Langmuir.* **2000**, 16, 5276–5283.
13. Clerc, M.; Laggner, P.; Levelut, A-M.; Rapp, G. *J. Phys. II (France).* **1995**, 5, 901–917.
14. Egelhaaf, S. U. *Curr. Opin. Colloid Interface Sci.* **1998**, 3, 608–613.
15. Egelhaaf, S. U.; Olsson, U.; Schurtenberger, P. *Physica B* **2000**, 276–278, 326–329.
16. Egelhaaf, S. U.; Schurtenberger, P. *Phys. Rev. Lett.* **1999**, 82, 2804.
17. Poon, W. C. K.; Renth, F. Evans, R. M. L.; Fairhurst, D. J.; Cates, M. E.; Pusey, P.N. *Phys. Rev. Lett.* **1999**, 83, 1239–1242.
18. Evans, R. M. L.; Poon, W. C. K.; Cates, M. E. *Europhys. Lett.* **1997**, 38, 595–600.
19. Poon, W. C. K.; Renth, F.; Evans R. M. L. *J. Phys.: Condens. Matter* **2000**, 12, A269–A274.
20. Buchanan, M.; Starrs, L.; Egelhaaf, S. U.; Cates, M. E. *Phys. Rev. E.* **2000**, 62, 6895–6905.
21. Stilbs, P. *Prog. Nucl. Magn. Reson. Spectrosc.* **1987**, 19, 1–45.
22. Veracini, C. A.; Struts, A. V.; Bezrukov, O. F. *Chem. Phys. Lett.* **1996**, 263, 228–234.
23. Lindblom, G.; Filfors, L. *Biochim. Biophys. Acta* **1989**, 988, 221–256.
24. Rilfors, L.; Eriksson, P-O.; Arvidson, G.; Lindblom, G. *Biochem. J.* **1986**, 25, 7702–7711.
25. Panitz, J. C.; Gradzielski, H.; Hoffmann, H.; Wokaun, A. *J. Phys. Chem.* **1994**, 98, 6812–6817.
26. Sein, A. PhD thesis, University of Groningen, 1995.
27. Brochard, F.; de Gennes, P. G. *Pramana. Suppl.* **1975**, 1, 1–21
28. Nallet, F.; Roux, D.; Prost, J. *J. Phys. II (France).* **1989**, 50, 3147–3165.
29. Lubensky, T. C.; Prost, J.; Ramaswamy, S. *J. Phys. II (France).* **1990**, 51, 933–943.
30. Sallen, L.; Oswald, P.; Sotta, P. *J. Phys. II (France)* **1997**, 7, 107–138.
31. Mullins, W. W.; Sekerka, R. F. *J. Appl. Phys.* **1964**, 35, 444–451.
32. Stevenson, D. G. *J. Textile Institute Trans.* **1953**, 44, 12-35.

33. Harker, R. P. *J. Textile Institute Trans.* **1959**, 50, 189.
34. Lawrence, A. S. C. *Faraday Discuss.* **1958**, 25, 51-58.
35. Lawrence, A. S. C. *Nature (London)* **1959**, 183, 1491-1494.
36. Sakurai, I.; Kawamura, Y. *Biochim. Biophys. Acta* **1984**, 777, 347–351.
37. Mishima, K.; Yoshiyama, K. *Biochim. Biophys. Acta* **1987**, 904, 149–153.
38. Sakurai, I.; Suzuki, T.; Sakurai, S. *Biochim. Biophys. Acta* **1989**, 985, 101–105.
39. Sakurai, I.; Suzuki, T.; Sakurai, S. *Mol. Cryst. Liq. Cryst. Sci. Technol., Sect. A* **1990**, 180, 305–311.
40. Sakurai, I.; Kawamura, Y.; Sakurai, T.; Ikegami, A.; Seto, T. *Mol. Cryst. Liq. Cryst. Sci. Technol., Sect. A* **1985**, 130, 203–222.
41. Sein, A.; Engberts, J. B. F. N. *Langmuir* **1996**, 12, 2913–2923.
42. Sein, A.; Engberts, J. B. F. N. *Langmuir* **1996**, 12, 2924–2931.
43. Buchanan, M.; Leng, J.; Egelhaaf, S. U.; Cates, M. E. *Polym. Prepr.* **2002**, 43, 860-861.
44. Haran, M.; Chowdhury, A.; Manohar, C.; Bellare, J. *Colloids Surf., A* **2002**, 205, 21–30.
45. Frette, V.; Tsafir, I.; Guedeau-Boudeville, M-A.; Jullien, L.; Kandal, D. Stavans, J. *Phys. Rev. Lett.* **1999**, 83, 2465–2468.
46. Santangelo, C. D.; Pincus, P. *Phys. Rev. E.* **2002**, 66, 061501
47. Buchanan, M. PhD thesis, The University of Edinburgh, 1999.
48. Buchanan, M.; Arrault, J.; Cates, M. E. *Langmuir* **1998**, 14, 7371–7377.
49. Diamant, H. PhD thesis, Tel-Aviv University, 1999.
50. Tanaka, T; Sun, S-T, Hirokawa, Y.; Katayama, S.; Kucera, J.; Hirose, Y.; Amiya, T. *Nature (London)* **1987**, 325, 796-798.
51. Virchow, R. *Virchow's Archive* **1854**, 6, 562–564.

Chapter 19

Droplet Microstructure and String Stability in Sheared Emulsions: Role of Finite-Size Effects

Jai A. Pathak*, Erin Robertson, Steven D. Hudson, and Kalman B. Migler*

Polymers Division, National Institute of Standards and Technology, Gaithersburg, MD 20899–8544

We discuss the influence of confinement on the microstructure of emulsions in steady shear flow and present initial results on experiments on stability of confined strings. We use flow visualization to get information about the structure and velocity of droplets and strings. In experiments on model emulsions comprising polyisobutylene (PIB) and poly(dimethylsiloxane) (PDMS), using a well-defined step-down protocol in shear, we find the organization of droplets in layers, starting with two layers at high shear rates and then one layer at lower shear rates. We reiterate arguments on the physics behind separation of droplets in layers. Strings formed under confinement can remain stable upon an increase in shear rate, so long as confinement effects due to the walls are still felt. Upon reduction in shear, strings turn into ellipsoidal droplets when interfacial stress effects overwhelm shear stress. Direct observation helps in unambiguous identification of string breakup and relaxation mechanisms. Evidence points to the role of confinement in "kinetic stabilization" of string breakup upon cessation of shear.

Introduction

The study of emulsion rheology was pioneered by Geoffrey Taylor ($1,2$), who not only experimentally identified the dimensionless groups (capillary number and viscosity ratio) that control droplet deformation in an emulsion in simple shear and hyperbolic flow fields, but also proposed a linear theory for droplet deformation in flow. The *droplet* Capillary number is defined as $Ca_{dr} = \eta_m \dot{\gamma} a / \sigma$ where η_m denotes the matrix viscosity, $\dot{\gamma}$ is the shear rate, a is the quiescent droplet radius and σ is the interfacial tension. The viscosity ratio is defined as $p = \dfrac{\eta_d}{\eta_m}$, where η_d is the droplet viscosity. Most work in the emulsion rheology literature has focused on the *bulk* case, where the droplet diameter, $2a$, is much smaller than the characteristic distance between shearing surfaces, e.g., the gap between parallel platens (d). In a recent study, Migler (3) has reported new physics in the confined regime ($2a \approx d$), where emulsion droplets coalesce with each other to form strings.

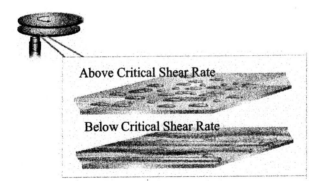

Figure 1. The droplet-string transition, viewed through a microscope. When an emulsion is sheared between parallel platens, droplets of the suspended phase (top) coalesce to form strings (bottom), with decreasing shear rate.

(See seventh page in color insert.)

The morphological droplet-string transition is summarized in Figure 1. String formation is governed not only by simultaneous droplet coalescence and breakup, typical in concentrated emulsions, but also by finite-size effects. As emulsions are sheared, smaller droplets coalesce with each other, and when the size of a droplet corresponds to critical capillary number conditions for that drop in the concentrated mixture (4), droplet breakup results. At any given shear rate, coalescence and breakup eventually reach a steady state, where the droplet size distribution no longer changes with time. The average droplet size grows with decreasing shear rate. Migler has shown that the droplet-string transition occurs at a "critical" shear rate (cf. Figure 1) when the droplet size in the vorticity direction, B, $\approx 0.5d$.

Results and Discussions

All details of the pure components' relative molecular mass and rheology, and experimental setup of the shear cell are in reference (5). PIB and PDMS samples behave as Newtonian liquids under the experimental conditions here with shear viscosities of 10 Pa·s each (25 °C). The gap width between parallel plates in the Linkam CSS-450 [1] shear cell was consistently set to 36 μm (standard uncertainty ± 3 μm; verified optically by microscope stage translation), with all observations in the vorticity-flow plane.

We have used a well-defined step-down shear protocol (5) to investigate droplet microstructure. In Figure 2, we present typical results of droplet velocimetry on a 10 % mass fraction PDMS/PIB emulsion (all compositions henceforth refer to mass fraction of PDMS). At $\dot{\gamma} = 6.75$ s^{-1} the histogram is distinctly bimodal, signifying that the droplets are moving in two layers and that there are two peaks in the concentration profile. Upon decreasing the shear rate, the two layer microstructure changes to one layer. At $\dot{\gamma} = 3.5$ s^{-1}, there is one sharp mode in the velocity histogram, located roughly halfway between the plates. The layering and overall composition profile is controlled by the interplay of droplet collisions (known, both experimentally (6) and from numerical simulations (7), to cause separation of droplet centers in the velocity gradient

---

[1] Certain commercial materials and equipment are identified in this paper in order to adequately specify the experimental procedure. In no case does such identification imply recommendation or endorsement by the National Institute of Standards and Technology, nor does it imply that these are necessarily the best available for the purpose.

direction), droplet migration (*8-10*) towards the centerline which arise from wall effects (*11-13*), and droplet packing constraints (*5*). Due to confinement, droplet self-diffusion (*14*) and gradient diffusion (*15*) in the velocity gradient direction may be neglected. In bulk systems, gradient diffusion and wall migration determine the composition profile (*16,17*). At higher shear rates, droplet collisions occur on shorter timescales than wall migration, and so they are effective in displacing droplets towards the walls and form a two-layer state. At lower shear rates, wall migration and collisions occur on comparable timescales, leading to accumulation of droplets in one layer.

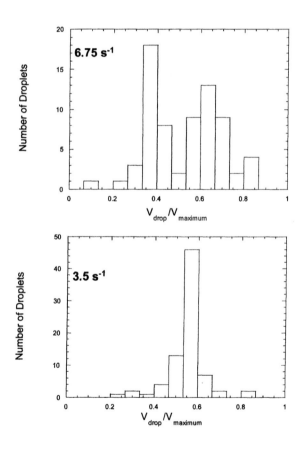

Figure 2. Typical histograms of the droplet velocity (v_{drop}) distribution in a 10 % PDMS/PIB emulsion at (a) $\dot{\gamma}$ = 6.75 s^{-1} and (b) 3.5 s^{-1}.

We have studied the effects of mixture composition on droplet microstructure (5), and summarized these results in the form of a rich morphology diagram (Figure 3) in the parameter space of mass fraction and shear rate. Formation of strings of the suspended phase was observed over a broad composition window. At each composition, experiments were stopped when strings were formed. We also found a non-transient morphology where we saw arrangement of the droplets in ordered pearl-necklace chain structures.

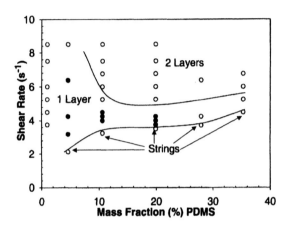

Figure 3. Microstructure in confined PIB/PDMS emulsions, during step-down in shear (d = 36 μm; p = 1). Points are experimental data and smooth curves only guide the eye. Filled black points denote ordered pearl-necklace droplet chains.

Armed with a broad view of droplet microstructure in confined emulsions, we now raise and attempt to answer some fundamental questions about the stability of confined strings. What stabilizes these strings? Is flow sufficient to stabilize these strings, or, must we invoke other physics to account for the stability? We consider two distinct cases. First, we turn the flow off completely, and observe string response. Second, we either increase or decrease the shear rate on flowing strings. The microstructure study (5) involved step-down in shear rate by small decrements, while in this stability study we make strings by directly going to the shear rate where strings were first seen to form. Strings form in a 10 % PDMS emulsion at $\dot{\gamma} = 3$ s^{-1} (cf. Figure 3). We make strings for the stability study by simply shearing the emulsion at $\dot{\gamma} = 3$ s^{-1}, instead of using the elaborate protocol in the microstructure study.

Upon cessation of flow, well-known interfacial tension driven flow instabilities set in and cause catastrophic breakup of the strings into droplets. For a "moderately long" string ($L/2R_o = 32$; R_o is the initial string radius; $2R_o/d = 0.70$), end-pinching (18) is the only instability seen: the ends of the original string first pinch off to form two drops and a daughter fragment, which undergoes further end-pinching to form droplets. The first end-pinching event on the original string is noticeably slower than the end-pinching event on the daughter fragment. This is evident in Figure 4 where we plot the time-dependent contour (end-to-end) length of the string, $L(t)$, normalized by the initial length of the string when the flow is turned off, $L(0)$, versus time (rendered dimensionless by the interfacial tension timescale $t^* = R_o \eta_m/\sigma$). This observation is consistent with that of Stone et al. (18), who found that in a multiple end-pinching sequence, the first pinch off is the slowest, and all subsequent pinch-offs are faster. The complete sequence of breakup of this string is shown in Figure 5.

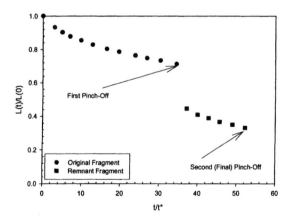

Cessation of Shear: 10 % PDMS T-41/PIB-800 Blend
End-to-End Length of String vs. Time During
Multiple End-Pinching Sequence

Figure 4. Time dependence of contour length of an end-pinching string in a 10% PDMS/PIB emulsion.

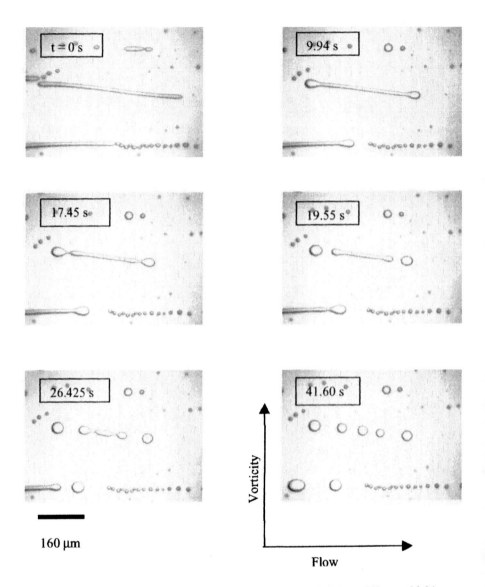

160 μm

*Figure 5. Relaxation of a moderately long string (L/2R₀ = 32) in a 10 %
PDMS/PIB emulsion upon cessation of shear.*

243

Cessation of shear in a longer string ($L/2R_o$ > 50; $2R_o/d$ = 0.79) results in string breakup by both end-pinching and the Rayleigh-Taylor instability (*19-21*)(cf. Figure 6). This observation is also consistent with the findings of Stone *et al.* (*18*) End-pinching is not visible in the lowermost string in Figure 5 as the ends of that string were not in the field of view of the video camera. However, the presence of end-pinching during the breakup of this string was confirmed by observing through the microscope objective, which has a much broader field of view than the camera, and afforded a clear view of one end of the string. A notable feature is that several tiny satellite droplets are formed during the breakup of strings into daughter droplets. For the lowermost string in Figure 5, 4 satellite droplets are seen in addition to the 5 daughter drops produced from the breakup of the string segment at the very bottom of Figure 5. The formation of satellite droplets signifies strong non-linearity of the late stages of breakup (*22-24*)!

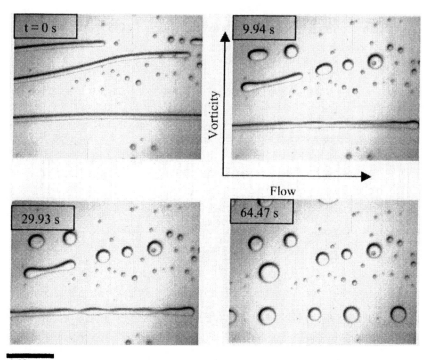

160 µm

Figure 6. Relaxation of a longer string ($L/2R_o$ > 50) in a 10% PDMS/PIB emulsion upon cessation of shear.

For the Rayleigh-Taylor instability (cf. Figure 6), the wavelength of the dominant mode ($\lambda = 165 \pm 5$ µm) which causes string breakup is $\approx 17\%$ larger than the Tomotika (20) prediction: $\lambda(p = 1) = 5.53 \times 2R_o$. $2R_o = 25.3$ µm gives the value of $\lambda = 140$ µm. The wavenumber, $\kappa = 2\pi R_o/\lambda$, of the dominant mode is lower than the Tomotika prediction. This issue, which may be a consequence of confinement, is being carefully studied in our group by experiments, as well as Lattice-Boltzmann simulations. We are specifically investigating whether any further increase in confinement causes further increase in wavelength. In addition, Tomotika's estimate of the breakup time of the thread (assuming the disturbances grow exponentially in time) is $t_b = (1/q)ln[\alpha_b/\alpha_o]$. Here q [$= \sigma\Omega(\lambda,p)/2\eta_m R_o$] is the growth rate, $\Omega(\lambda,p)$ is the dimensionless disturbance growth rate (20), $\alpha_o = [21kT/8\pi^{3/2}\sigma]^{1/2}$ is the Brownian disturbance amplitude (25) and α_b is the amplitude at breakup (which equals the average string radius). Using $R_o = 12.65$ µm at $T = 298$ K, and $\Omega = 0.1$ (20) yields $t_b = 8.3$ s, significantly smaller than experimental $t_b = 52$ s for the data in Figure 6. This t_b estimate is valid strictly when capillary instability is the only breakup mechanism. End-pinching occurs on shorter timescales (18) than the Rayleigh-Taylor instability. For a sufficiently long string breakup begins by end-pinching and capillary waves later appear on the middle section. Evidently, confinement causes a "kinetic stabilization" (26) of the breakup of these strings. Breakup is slowed down over the bulk case because fluctuations in the velocity gradient direction are suppressed due to significant hydrodynamic interactions between the walls and the string.

We now discuss the case where the shear rate is increased in small steps. At each new shear rate we observe the response of the strings for at least half an hour, before increasing the shear rate to the next higher value. During the step up, the average string velocity increases, as expected, and the strings remain *centered* between the plates, evident in Figure 7, where the velocity of the strings (in dimensional and dimensionless forms) is plotted versus shear rate.

When the string velocity is normalized by the velocity of the steadily-rotating plate, $v_{max} = \dot{\gamma} d$, we find that v/v_{max} equals 0.5, within experimental error, at each increasing shear rate, signifying that the strings are centered on the centerline between the parallel plates. In the starting state ($\dot{\gamma} = 3$ s^{-1}), strings are already centered between the plates because wall-migration overcomes droplet collisions in the one layer state before strings are formed. We remind the reader that collisions are essentially arrested (5) in the one layer state, whose formation precedes the formation of strings. Due to the appreciable droplet size, wall migration effects become strong and send droplets to the center. Once in the center, the droplets form pearl necklaces and ultimately coalesce to form strings, which are also on the centerline.

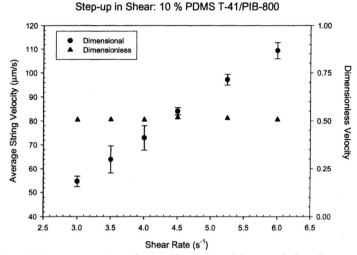

Figure 7. Average string velocity (dimensional form on left ordinate; dimensionless form on right ordinate) in a 10 % PDMS/PIB emulsion. Bars on data points denote standard uncertainties equal to the standard deviation.

A natural consequence of the increase in shear rate is that the strings are elongated in the flow direction, causing contraction in the velocity gradient and vorticity directions, to satisfy volume conservation. We quantify the string diameter in the vorticity direction, and track its decrease with increasing shear rate (cf. Figure 8). We assume that the strings are axisymmetric, implying that the diameters in the velocity gradient and vorticity directions are equal. This assumption is justified in the case of *increasing* shear, as the strings thin out upon increase in shear and feel progressively less influence of the walls; this assumption cannot be made when the walls squash the strings, a situation likely to arise from string-string coalescence which increases string thickness in the velocity gradient direction and causes increased hydrodynamic interactions between the strings and the walls (*27*).

Simultaneous observation of the strings and quantification of the string dimensions in the shear flow field offer insight into the physics behind the stability of strings upon increasing shear. We find that the strings survive with increasing shear rate until the diameter in the vorticity direction (assumed equal to the diameter in the velocity gradient direction) decreases to about half the gap width. Upon further decrease in the vorticity diameter, the strings become unstable and start to break up. Breakup (facilitated by droplet-string and string-

string collisions) becomes *clearly visible* (aspect ratio of strings plummets) at $\dot{\gamma} = 6.0 \text{ s}^{-1}$, where the string diameter $2R$ is 15 µm, defining a critical capillary number $Ca_{cr,string} = 0.18$ for the strings (σ for the PDMS/PIB system is 2.5×10^{-3} N/m (*28*); the *string* capillary number is defined as $Ca_{string} = \eta_m \dot{\gamma} R / \sigma$). This observation strongly suggests that the stabilizing influence of the walls (which induce string formation) is necessary but not sufficient for string stability. The observation of wall-induced stabilization is also consistent with the findings of Son *et al.* (*29*), who have experimentally studied confinement effects on the capillary instability of a polymer thread in a quiescent polymeric matrix. They also observed that when the thread diameter becomes greater than or equal to half the matrix height, confinement effects on the capillary instability are felt in terms of simultaneous decrease in the growth rate of the instability and an increase in the wavelength of the instability (also observed here). Close proximity of the walls is necessary for the stability of the strings, but not sufficient because even when the strings are wall-stabilized, flow is also necessary to keep them intact. When flow is turned off completely, interfacial tension drives instabilities that eventually break up the strings into droplets. Herein lies the complex interplay of flow and confinement in string stabilization.

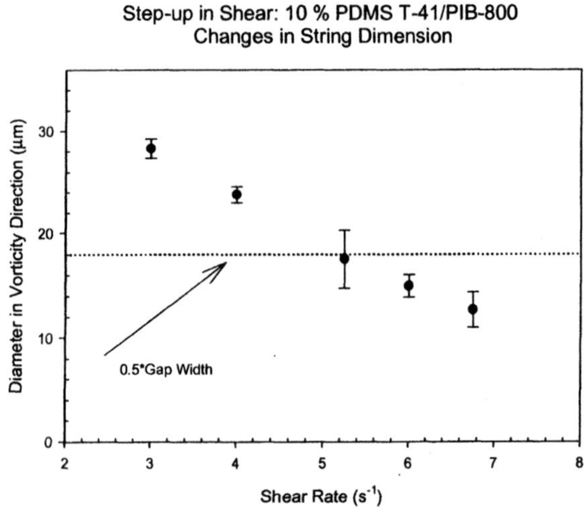

Figure 8. Average string diameter in vorticity direction in a 10 % PDMS/PIB emulsion. Dotted line signifies half the gap width (18 µm). Bars on data points denote standard uncertainties equal to the standard deviation.

Upon forming strings at $\dot{\gamma} = 3$ s^{-1} and then reducing the shear rate in small decrements of 0.5 s^{-1}, we find that the ends of the strings gradually retract towards the center, due to increasing importance of interfacial stress over the deforming shear stress. All the strings finally turn into ellipsoidal droplets at $\dot{\gamma} = 1.5$ s^{-1} with an average $L/B = 1.65 \pm 0.23$ (L and B denote the major and minor axes of the ellipsoid, respectively), when interfacial stress completely overcomes shear stress.

When we combine the results of step-up and step-down in shear, we find that there is a definite "window" in shear rate across which confined strings are stable. For the 10% PDMS/PIB emulsion, strings formed at 3 s^{-1} are stable up to $\dot{\gamma} = 5.25$ s^{-1} during step-up in shear and about 2.0 s^{-1} for step-down. We also find a *hysteresis* in morphology upon comparison of our present results with those presented in the morphology diagram (cf. Figure 3). During step down in shear, the one layer state is seen between 4 s^{-1} and 6 s^{-1}, but during step-up in shear, *strings* are seen to persist over the same shear rate interval. We have also checked for any possible path dependence to the morphology, i.e., whether the morphology observed depends on whether we directly increase $\dot{\gamma}$ from 3.0 s^{-1} to 5.25 s^{-1}, or increase $\dot{\gamma}$ in small steps of 0.5 s^{-1} or 0.75 s^{-1}. The final morphology at 5.25 s^{-1} is independent of the path taken. Similar results are found for step-down, where it is seen that the strings always relax to droplets, irrespective of whether we decrease the shear rate in small decrements to 1.5 s^{-1} or go directly from 3.0 s^{-1} to 1.5 s^{-1}.

Conclusions

A definite consequence of confinement is the organization of droplets in layers. This result is in sharp contrast to the findings of King and Leighton (*16*), who see *only one peak* in the volume fraction profile, under all conditions. The a/d ratio for their system is $\ll 1$, as discerned from Table 1 in ref. (*16*), strongly suggesting that droplet layering is a finite-size effect, as gravitational effects are negligible here (*5*). The interplay of confinement and flow governs the survival and stability of strings over a certain shear rate window upon increase in shear. Evidence for the role of confinement is adduced by comparison of the string dimension with the gap width. When the string diameter is greater than or equal to half the gap width, the strings survive upon step-up in shear. Additional evidence for the role of confinement is provided by results from recent and earlier experiments (*3*) in our group, where we find that strings whose vorticity direction thickness is equal to the gap width survive over much longer timescales than thinner strings. Interfacial tensions eventually wins at low shear rates, as expected, and the strings turn into droplets. We are currently focusing on experimentally investigating the effects of string-string interactions (which gain prominence in more concentrated emulsions) on string stability and the simultaneous breakup of multiple strings. In addition, as pointed out earlier, we are looking at how imposing more severe confinement changes the characteristic wavelength of the dominant mode of the Rayleigh-Taylor instability and the timescale of string breakup.

Acknowledgments

We thank Alex Jamieson (CWRU) for use of his shear cell. ER was a NIST Summer Undergraduate Research Fellow. We thank Jack Douglas and Nicos Martys for comments, and Kathy Flynn and Melissa Davis for assistance.

References

1. G.I.Taylor *Proc.R.Soc.Lond.A* **1932,** *138* 41-48.

2. G.I.Taylor *Proc.R.Soc.Lond.A* **1934,** *146* 501-523.

3. K.B.Migler *Phys.Rev.Lett.* **2001,** *86* 1023-1026.

4. K.M.B.Jansen; W.G.M Agterof; J.Mellema *J.Rheol.* **2001,** *45* 227-236.

5. J.A.Pathak; M.C.Davis; S.D.Hudson; K.B.Migler *J.Coll.Interface Sci.* **2002,** *255* 391-402.

6. S.Guido; M.Simeone *J.Fluid Mech.* **1998,** *357* 1-20.

7. M.Loewenberg; E.J.Hinch *J.Fluid Mech.* **1997,** *338* 299-315.

8. A.Karnis; S.G.Mason *J.Coll.Interface Sci.* **1967,** *24* 164-169.

9. P.C.-H.Chan; L.G.Leal *Int.J.Multiphase Flow* **1981,** *7* 83-99.

10. J.R.Smart; D.T.Leighton *Phys.Fluids A* **1991,** *3* 21-28.

11. C.E.Chaffey; H.Brenner; S.G.Mason *Rheologica Acta* **1965**, *4* 64-72.

12. C.E.Chaffey; H.Brenner; S.G.Mason *Rheologica Acta* **1967**, *6* 100.

13. P.C.-H.Chan; L.G.Leal *J.Fluid Mech.* **1979**, *92* 131-170.

14. M.Loewenberg; E.J.Hinch *J.Fluid Mech.* **1996**, *321* 395-419.

15. F.R.da Cunha; E.J.Hinch *J.Fluid Mech.* **1996**, *309* 211-223.

16. M.R.King; D.T.Leighton *Phys.Fluids* **2001**, *13* 397-406.

17. B.E.Burkhart; P.V.Gopalakrishnan; S.D.Hudson; A.M.Jamieson;

    M.Rother; R.H.Davis *Phys.Rev.Lett.* **2001**, *87* 098304-098308.

18. H.A.Stone; B.J.Bentley; L.G.Leal *J.Fluid Mech.* **1986**, *173* 131-158.

19. Lord Rayleigh *Phil.Mag.* **1892**, *34* 145-154.

20. S.Tomotika *Proc.R.Soc.Lond.A* **1935**, *150* 322-337.

21. Lord Rayleigh *Proc.Lond.Math.Soc.* **1878**, *10* 4-13.

22. D.W.Bousfield; R.Keunings; G.Marrucci; M.M.Denn *J.Non-Newtonian Fluid Mech.* **1986**, *21* 79-97.

23. M.Tjahjadi; H.A.Stone; J.M.Ottino *J.Fluid Mech.* **1992**, *243* 297-317.

24. M.Tjahjadi; J.M.Ottino; H.A.Stone *AIChE J.* **1994**, *40* 385-394.

25. W.Kuhn *Kolloid Zeitschrift* **2002**, *132* 84-99.

26. J.G.Hagedorn; N.S.Martys; J.F.Douglas *Phys.Rev.E* **2002**, *Submitted.*

27. N.S.Martys; J.F.Douglas *Phys.Rev.E* **2002**, *63* 031205-1-031205-18.

28. M.Wagner; B.A.Wolf *Macromolecules* **1993**, *26* 6498-6502.

29. Y.-G.Son; N.S.Martys; K.B.Migler *Macromolecules* **2002**, *Submitted.*

Chapter 20

Viscous Fingering of Silica Suspensions Dispersed in Polymer Fluids

Masami Kawaguchi

Department of Chemistry for Materials, Faculty of Engineering,
Mie University, Mie 514–8505, Japan

We make an experimental study of the viscous fingering of shear thinning silica suspensions in a radial Hele-Shaw cell and shear thickening silica suspensions in a linear Hele-Shaw cell injected by air. For the shear thinning silica suspensions, the viscous fingering instability is strongly related to the polymer concentration in the dispersant rather than the silica concentration. For the shear thickening silica suspensions, the imposed shear rate at which the viscous fingering instability is observed for the first time, is close to the critical shear rate of the corresponding shear thickening silica suspensions. The finger velocities of the shear thinning silica suspensions are in agreement with the modified Darcy's law, where the constant viscosity is replaced by the shear rate dependent viscosity. On the other hand, the finger velocities of the shear thickening silica suspensions with the silica concentrations larger than 7.5 wt % are much lower than the prediction of the modified Darcy's law.

Introduction

Motion of a fluid produces a variety of spatio-temporal shapes (pattern) in nature: flowing clouds with changes in their shapes, traveling an air bubble in a bottle filled with water, and burning flames. On the other hand, in a quasi-two dimensional space, one of the simplest problems of fluid motion is the Saffman-Taylor problem in which two fluids move in the narrow space between two plates (*1*), namely in a Hele-Shaw cell (*2*). Pressure gradient driven pattern formation of an interface between two fluids in Hele-Shaw cells, viscous fingering occurs when a more viscous fluid is displaced by a less viscous fluid injected. Viscous fingering produces pattern formation far from equilibrium, it is a model system for flow through porous media, and it can be related to the recovery of crude oils in oil wells as well as the blowing process of plastic products. Viscous fingering in Newtonian fluids has been theoretically and experimentally well understood and the developments of the viscous fingering in Newtonian fluids have been deeply reviewed (*3-5*). On the other hand, during recent two decades several research groups have paid much attentions to investigate the viscous fingering non-Newtonian fluids (*3-7*), such as polymer solutions, liquid crystals, forms, gels, and suspenstions, but our knowledge has been little in comparison with that of Newtonian ones due to their complex rheological properties.

In this study, we report viscous fingering experiments of two silica suspensions dispersed in polymer fluids: one is a silica suspension dispersed in aqueous hydroxylpropyl methyl cellulose (HPMC) solution and it shows shear thinning behavior; the other is a silica suspension dispersed in poly propylene glycol (PPG) and it indicates shear thickening response. We describe their viscous fingering instabilities in terms of changes in pattern morphology and finger pattern growth by taking into account the rheological responses of two silica suspensions.

Experimental

Materials

An HPMC sample kindly supplied by Shin-Estu Chemical Co. was purified by the method described previously (*8*) and its molecular weight was determined to be 250×10^3 by the intrinsic viscosity measurement. PPG with the molecular weight of 725 was purchased from the Aldrich Chemical Co. and it was used without further purification. Silica suspensions were prepared by mechanically mixed Aerosil 130 silica powder supplied from the Japan Aerosil Co. with aqueous HPMC solutions of different concentrations (HPMC-silica suspension) and with PPG (PPG-silica suspension). The silica concentrations in the former suspensions were fixed at 5.0, 7.5, and 10.0 wt %, whereas the silica contents of the latter suspensions were 2.5, 5.0, 7.5, and 10.0 wt %. Moreover, the concentrations of HPMC in the supernatants of the HPMC-silica suspensions should be lower than those in the prepared HPMC solutions, since HPMC chains are adsorbed on the silica surfaces (9).

Instrumentations

The viscous fingering experiments were performed at 25 °C using two Hele-Shaw cells: a radial cell made by using two plane-glass plates (0.8x50x35 cm^3) with a silicon wafer spacer of 0.05 cm thickness clamped in between the plates (10); a linear cell made of the same two glass plates as the radial cell, clamped along their sides with a U-shaped silicone rubber sheet (0.05x50x15 cm^3) and with silicon wafer spacers of 0.05 cm thickness in between the plates (11). In the linear cell, the width of the channel was fixed at 3 cm by using the rubber sheet. The HPMC-silica suspension was injected through the inlet at the center of the top plate for the radial cell to form a 10 cm sample radius, whereas for the linear cell the PPG-silica suspension was injected into the inlet at the distance L of 20 cm from the short unsealed edge of the top plate. Air was injected through the inlet at the fixed injection pressure of 5.0 kPa for the HPMC-silica suspensions, whereas for the PPG-silica suspensions the injection pressures were changed from 1.0 to 30 kPa. The generated patterns were recorded with a CCD camera-recorder method. The images of the recorded patterns were analyzed by a Himawari-60 digital image analyzer.

Rheological measurements of the silica suspensions were performed using a Paar Physica MCR300 rheometer with a cone-plate geometry.

Results and Discussion

HPMC-Silica Suspensions

HPMC chains adsorb onto the silica surface (9) and their adsorbed amounts are determined to be ca. 0.12g/g, irrespective of the silica concentration and HPMC. Figure 1 displays steady-state shear viscosities of the 5.0, 7.5, and 10.0 wt % silica suspensions dispersed in a 1.5g/100 mL HPMC solution, where the HPMC concentrations in the supernatants are 0.90, 0.60, and 0.30 g/100 mL in order of the silica content, as a function of the shear rate, together with that of 1.5 g/100 mL HPMC solution. The HPMC solution shows the existence of a Newtonian region followed by the weak shear thinning. The silica suspensions show shear-thinning behavior typical for aggregated suspensions. This is attributed to the mechanical strength of the aggregated silica suspensions, whose structure was a fractal-like, as determined by small-angle neutron scattering measurement (9).

Moreover, the HPMC-silica suspensions show solid-like viscoelastic responses: their dynamic storage moduli G' are larger than the dynamic loss moduli G" at small and linear strain ranges. On the other hand, the HPMC solution has much larger values of G" than G' in the frequency ranges from 0.1 to 100 rad/s due to liquid-like viscoelastic matter.

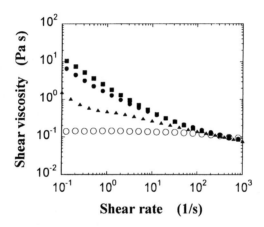

Figure 1. Double-logarithmic plots of steady-state shear viscosities as a function of shear rate for 1.5 g/100 mL HPMC solution (open circle), 5.0 (filled triangle), 7.5 (filled circle), and 10.0 wt % (filled square) silica suspensions dispersed in a 1.5 g/100 mL HPMC solution.

Figure 2 shows typical fingering patterns of the 5.0, 7.5, and 10.0 wt % silica suspensions dispersed in a 1.5 g/100 mL HPMC solution. The resulting fingering patterns change from side-branching and suppressed side-branching to tip-splitting patterns with an increase in the silica concentration, namely a decrease in the free HPMC concentration. From a comparison with fingering patterns of HPMC solutions (10,11), it is worth noting that the characteristics of the morphological changes in fingering patterns of the HPMC-silica suspensions are strongly related to the HPMC concentration in the supernatant.

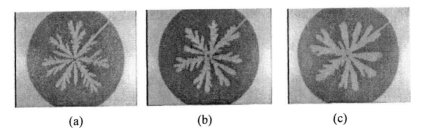

Figure 2. Typical fingering patterns formed by air (white) injected into the (a) 5.0, (b) 7.5, and (c) 10.0 wt % silica suspensions (dark).

We estimated the finger tip velocity v_t defined by the initial slope for a plot of a given distance L_f from the center of the cell against the necessary time for the outermost parts of the fingers to reach the distance. The L_f was nearly proportional to the time up to around the distance of 5 cm and above the distance the plot was gradually and upwardly deviated from the straight line. The v_t value should be related to an average finger velocity v of the fluid far away from the finger. The v value can be modified by the displaced ratio λ of the more viscous fluid by the less viscous one for the circular area of the fingering pattern at the distance of L_f = 5cm as follows: $v = v_t \lambda$. The value of v should be related to Darcy's law $v = (b^2/12\eta) \nabla p$, where b is the cell gap, η is the viscosity of the displaced and more viscous fluid, and ∇p is the pressure gradient. The ∇p is defined as the ratio of the injection pressure of 5.0 kPa and the distance of $L - L_f$ (= 5 cm), where L is the radius of the fingering pattern at L_f (= 5 cm).

Moreover, in order to verify Darcy's law for the HPMC-silica suspensions, we should replace the constant viscosity η by an effective viscosity η_{eff} at the imposed shear rate defined by $2v_t/b$ since the silica suspensions show non-Newtonian behavior as seen in Figure 1. Thus, the modified Darcy's law can be defined as $v = (b^2/12\eta_{eff}) \nabla p$ and the η_{eff} value can be obtained from the interpolation of a plot of the steady state viscosity against the shear rate as shown in Figure 1. In Figure 3, the values of v for the HPMC-silica suspensions are plotted against the value of $(b^2/12\eta_{eff}) \nabla p$ by taking account of nonlinear dynamics and the dashed straight line drawn in the figure corresponds to the value of one. The fit of the resulting data to the modified Darcy's law is good. The v values of the viscous fingering experiments in HPMC solutions with different concentrations, which showed shear thinning behavior, were in good agreement with the modified Darcy's law (12).

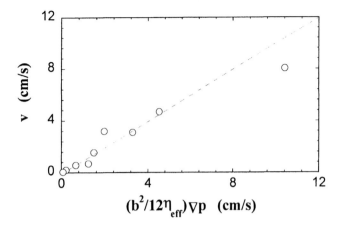

Figure 3. Plots of v for various HPMC-silica suspensions as a function of $(b^2/12\eta_{eff})\nabla p$ in the radial cell. The dashed straight line has a slope of one.

PPG-Silica Suspensions

Steady-state shear viscosities of the 2.5, 5.0, 7.5, and 10.0 wt % PPG-silica suspensions are displayed as a function of the shear rate in Figure 4, together with that of PPG. PPG shows Newtonian behavior, whereas the PPG-silica suspensions indicate stronger shear thickening with an increase in the silica concentration since the silica particles interact by hydrogen bonding to form a gel (*13*). The critical shear rate corresponding to the onset of steady shear thickening becomes lower with an increase in the silica concentration. Such a concentration dependence of the critical shear rate is similar to those reported previously (*14*).

Figures 5 and 6 show typical fingering patterns of PPG as well as the 2.5 and 7.5 wt % PPG-silica suspensions at the different injection pressures of 6.0 and 30 kPa, respectively. The fingering pattern of PPG changes from a stable finger (Saffman-Taylor finger (*1*)) at 6.0 kPa, whereas it is an unstable one, i.e. a tip-splitting pattern at 30 kPa, and below the injection pressure of 15 kPa PPG gives only the Saffman-Taylor finger. The Morphological transition of the fingering pattern from the Saffman-Taylor finger to an unstable one is called as the Saffman-Taylor or viscous fingering instability (*1*). The Saffman-Taylor instability for the PPG-silica suspensions occurs at a lower injection pressure than that for PPG, due to non-Newtonian behavior, namely shear thickening as shown in Figure 4.

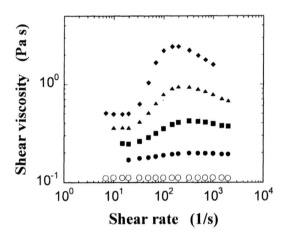

Figure 4. Double-logarithmic plots of steady-state shear viscosities as a function of shear rate for PPG (circular) as well as 2.5 (filled circular), 5.0 (filled square), 7.5 (filled triangle), and 10.0 (filled diamond) wt % silica suspensions in dispersed in PPG.

Figure 5. Typical fingering patterns for PPG as well as the 2.5 and 7.5 wt % silica suspensions dispersed in PPG at the injection pressure of 6.0 kPa from top to bottom.

*Figure 6. Typical fingering patterns for PPG as well as the 2.5 and 7.5 wt %
silica suspensions dispersed in PPG at the injection pressure of 30 kPa from top
to bottom.*

We notice some general trends for changes in the pattern growth: 1) at a
fixed injection pressure the resulting instability occurs at an early stage of the
pattern growth and the finger becomes wider with an increase in the silica
content; 2) with an increase in the injection pressure the finger becomes
narrower and a larger number of tip-splitting occur, however, at the injection
pressure of 30 kPa a very disrupted chaotic-like pattern in which no clear
branching cascade could be detected for weakly shear thinning clay suspensions
(*15*), was not observed, but weak front oscillations were observed.

The finger tip velocity v_t can be calculated using a plot of the distance L_f
from the inlet to the tip of the grown finger against the necessary time. In the
plot a straight portion is obtained for a distance $L_f = 5$-6 cm from the inlet and
beyond the distance the plot data are positively deviated from the initial straight
line, irrespective of the injection pressure or the sample. The relative finger
width λ defined by the ratio of the finger width and the cannel width W, can be
calculated from the increase of the displaced area A by air in time t: $\lambda = (1/$
$v_t)(dA/dt)/W$. Each plot of A against t has a straight portion for the distance L_f of
5-6 cm. Therefore, for the distance $L_f = 5$-6 cm the finger growth can be
regarded to be in the steady state.

In order to study the onset of the Saffman-Taylor instability for the PPG-
silica suspensions, we estimated the imposed shear rate $2v_t/b$ at which the
unstable finger for the first time appears. The $2v_t/b$ values are obtained as 72, 30,
17, and 11 s^{-1} for the 2.5, 5.0, 7.5, and 10.0 wt% PPG-silica suspensions,
respectively and they are independent of the injection pressure. Moreover, these
values are close to the critical shear rates of the corresponding PPG-silica

suspensions as shown in Figure 4.   The observed shear thickening behavior is due to the gel formation of the silica particles, leading to the Saffman-Taylor instability.   Thus, it is noteworthy that the onset of non-linear rheological response induces the instability of viscous fingering phenomena.   For the shear thinning polymer solutions (*16*) and Borger fluids, namely, constant viscosity elastic fluids (*17*), their viscous fingering instabilities were correlated with the onset of non-linear behavior of their rheological properties.

The relative finger width λ is well known related to the dimensionless parameter $1/B = 12$ ($\eta_{eff} v_t/\gamma$) $(W/b)^2$ (*18,19*), where γ is the surface tension of the displaced fluid.   At the present time the surface tensions of the silica suspensions are not available and the values of λ are plotted as a function of $1/B' = 12$ ($\eta_{eff} v_t$) $(W/b)^2$ in Figure 7.   The λ value of PPG gradually decreases and tends to approach to a plateau value of ca. 0.5 at higher $1/B'$ value. The λ values of the 2.5 and 5.0 wt % PPG-silica suspensions also decrease with an increase with $1/B'$ and they are smaller than that of PPG at the higher $1/B'$.

Moreover, the λ values of the 7.5 and 10.0 wt % silica suspensions at the lower value of $1/B'$ are larger than that of PPG.   The presence of the larger value of λ qualitatively agrees with the theoretical prediction of Ben Amar and Poire (*20*) for shear thickening fluids.

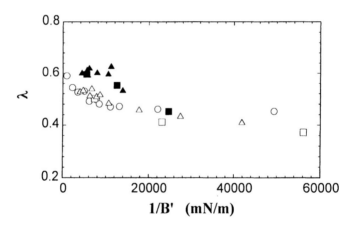

Figure 7.   Plots of λ for PPG (open circle) as well as 2.5 (open square), 5.0 (open triangle), 7.5 (filled square), and 10.0 (filled triangle) wt % silica suspensions in dispersed in PPG as a function of 1/B'.

The v values for steady state finger growth of PPG and PPG-silica suspensions are plotted against the value of $(b^2/12\eta_{eff})\,\nabla p$ in Figure 8, where ∇p is defined by the ratio of the injection pressure and the distance $L - L_f\,(= 5\text{-}6\text{ cm})$. The dashed straight line drawn in the figure corresponds to the value of one. The fit of the resulting data to the modified Darcy's law is relatively good for PPG and the 2.5 and 5.0 wt % silica suspensions. Thus, it is found that the finger growth of the weakly shear thickening silica suspensions is sufficient with the modified Darcy's law.

On the other hand, the v values of the 7.5 and 10.0 wt % silica suspensions are more negatively deviated from that predicted by the modified Darcy's law with an increase in the silica concentration. They tend to approach a plateau value with an increase in $(b^2/12\eta_{eff})\,\nabla p$. This means that the finger velocities of the stronger shear thickening silica suspensions cannot be interpreted by only the displacement of the viscosity with the shear dependent viscosity. At the present time, we do not have any explicit explanation of such smaller v values of the 7.5 and 10.0 wt % silica suspensions. The wetting layer remaining in the cell after the finger growth may occur the reduction of the cell gap, leading to the smaller finger velocity, and further experiments should be accumulated.

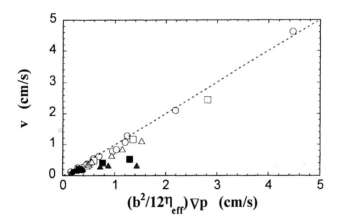

Figure 8.   Plots of v for PPG as well as 2.5, 5.0, 7.5, and 10.0 wt % silica suspensions in dispersed in PPG as a function of $(b^2/12\eta_{eff})\,\nabla p$.   The dashed straight line has a slope of one.   Symbols are the same as in Figure 7.

Conclusions

Changes in the viscous fingering patterns of HPMC-silica suspensions are mainly attributed to the free concentration of HPMC. For the PPG-silica suspensions, the onset of the Saffman-Taylor instability is strongly related to the critical shear rate of the corresponding silica suspensions. Except for the stronger shear thickening PPG-silica suspensions, the value of v for the silica suspensions showing shear thinning or shear thickening behavior is in good agreement with the modified Darcy's law, irrespective of non-Newtonian response.

Acknowledgements

This work was partially supported by Grant-in-Aid for Scientific Research on Priority Area (A) "Dynamical Control of Strongly Correlated Soft Materials" (No. 413/13031045) from the Ministry of Education, Science, Sports, Culture, and Technology in Japan.

References

1. Saffman, P. G.; Taylor, G. I. *Proc. Royal Soc. Lond. A*, **1958**, 245, 312.
2. Hele-Shaw, H. J. S. *Nature*, **1898**, 58, 34.
3. Vicek, T., *Fractal Growth Phenomena*, World Scientific: Singapore, 1992.
4. Meakin, P., Fractal, Scaling and Growth Far from Equilibrium; Cambridge Univ. Press: Cambridge, 1998.
5. NcCloud, K. V.; Maher, J. V. *Phys. Rep.* **1995**, 260, 139.
6. Fast, P.; Kondic, L.; Shelly, M. J.; Palffy-Muhoray, P. *Phys. Fluids* **2001**, 13, 1191.
7. Kawaguchi, M. *Recent. Res. Devel. Polymer Sci.* **2002**, 6, 139 and references therein..
8. Kato, T.; Tokuya, T.; Takahashi, A. *Kobunshi Ronbunshu* **1982**, 39, 293.
9. Kawaguchi, M.; Kimura, Y.; Tanahashi, T.; Takaeoka, J.; Kato, T.; Suzuki, J.; Funahashi, S. *Langmuir* **1995**, 11, 563.
10. Makino, K.; Kawaguchi, M.; Aoyama, K.; Kato, T. *Phys. Fluids* **1995**, 7, 455.
11. Kawaguchi, M.; Shibata, A.; Shimomoto, K.; Kato, T. *Phys. Rev. E.* **1998**, 58, 785.
12. Kawaguchi, M. *Nonlinear Analysis* **2000**, 47, 907.
13. Raghavan, S. R.; Walls, H. J.; Khan, S. A. *Langmuir* **2000**, 16, 7920.

14. Van Egmond, J. W. Curr. Opin. Colloid Interface Sci. **1998**, 3, 385 and references therein.
15. Van Damme, H.; Lemaire, E. Disorder and Facture; Charmet, J. C.; Roux, S.; Guyon, E., Eds.; Plenum Press: New York, 1990; p 83.
16. Kawaguchi, M.; Hibino, Y.; Kato, T. *Phys. Rev. E.* **2001**, 64, 051806.
17. Vlad, D. H.; Maher, J. V. *Phys. Rev. E.* **2000**, 61, 5439.
18. Bensimon, D.; Kadanoff, L. P.; Liang, S.; Shariman, B. I.; Tang, C. *Rev. Mod. Phys.* **1986**, 58, 977.
19. Tabeling, P.; Zocchi, G.; Libchaber, A. *J. Fluid. Mech.* **1987**, 177, 67.
20. Ben Amar, M.; Poire, E. C. *Phys. Fluids* **1999**, 11, 1757.

Phase Separation

Chapter 21

Controlled Pattern Formation in Some Block Copolymer Systems

Yoav Tsori

ESPCI, 10 Rue Vauquelin, Paris 75231, Cedex 05, France

We discuss several mechanisms useful in control of pattern formation in copolymeric systems. Chemically patterned substrate is shown to induce ordering in confined diblock copolymer melt. The strength and range of this ordering depend on interfacial interactions and surface feature size. Electric field is effective in aligning a sample in a desired direction by the ``dielectric mechanism''. We explain how this effect can be exploited in certain situations. We consider the new effect of dissociated mobile ions in a melt in electric field. Orienting forces in these non-equilibrium systems are calculated and are found to be important to alignment. Some morphological changes predicted to occur are illustrated.

Many systems in nature show fascinating and complicated pattern formation. Example include diblock copolymer melts (*1,2*), mixtures of diblock and homopolymers, aqueous solutions of lipids or surfactants, Langmuir monolayers, magnetic garnet films (*3*), chemical reactions (*4*), biological systems (*5*) and convective rolls in heated water (*6*). The modulated phases in many of these systems result from a competition between short- and long-range forces. Here we consider as a model system confined block copolymers (BCP) with or without electric field. There is a similarity between morphologies in BCPs and the Turing patterns observed in inorganic acidic chlorite-iodide-malonic reactions

(*4*). In the latter there is long-range inhibition and short-range activation. In the BCP system the competition is between chain stretching and short-range repulsion.

BCPs are composed of two or more chemically distinct chains, or blocks, joined by a covalent bond. In the low temperature regime, the macro phase-separation occurring in most polymers is inhibited because of chain connectivity, giving rise to mesoscopically ordered phases depending on chain length, architecture, etc. The most simple BCP is a linear diblock copolymer comprised of two connected linear chains. The equilibrium phase diagram has been extensively studied and found to consist of a disordered phase (two chains mixed) in the high temperature regime, and Lamella, Cylinder, Gyroid and Sphere phases below the critical temperature. In the next section we discuss first equilibrium systems, and show how morphologies are affected when the melt is put in contact with confining surfaces. Intrinsically non-equilibrium systems in electric fields are considered next, and orientation by electric field is explained. We show that the presence of dissociated ions in the melt can lead to strong aligning forces (accompanied by heating) in AC electric field.

Surface Induced Ordering in a BCP Melt Above ODT

The A/B diblock copolymer melt is characterized by two parameters: $f=N_A/N$ and χN, where N_A is the number of A monomers in a chain of $N=N_A+N_B$ monomers, and $\chi \sim 1/T$ is the so-called Flory parameter characterizing the repulsion between the chains. The bulk free energy of the system

$$F_b / kT = \int \left\{ \frac{1}{2}\tau\varphi^2 + \frac{1}{2}h\left(q_0^2\varphi + \nabla^2\varphi\right)^2 + \frac{1}{6}\Lambda\varphi^3 + \frac{1}{24}\Delta\varphi^4 \right\} d^3r \qquad (1)$$

is given as a function of the order parameter $\varphi(r)=\varphi_A(r)-f$, where $\varphi_A(r)$ is the local A-monomer density. $d_0=2\pi/q_0$ is the fundamental periodicity in the system, and is expressed by the polymer radius of gyration R_g, through $q_0=1.95/R_g$. In addition, $\tau=2\rho N(\chi_c-\chi)$ and $h=1.5\rho c^2 R_g^2/q_0^2$ (*7*). The second length scale in the system is determined by the ratio of two parameters, $(\tau/h)^{1/4} \sim (N\chi_c-N\chi)^{1/4}$, and characterizes the decay of surface-induced modulations. The Flory parameter χ measures the distance from the Order-Disorder Temperature (ODT), having the value $\chi_c \sim 10.49/N$. Finally, $\rho=1/Na^3$ is the chain density per unit volume, and Λ and Δ are the three- and four-point vertex function calculated by Leibler (*1,7*).

This free energy describes a system in the disordered phase having a uniform $\varphi=0$ for $\chi<\chi_c$, while for $\chi>\chi_c$ and for symmetric $(f=\frac{1}{2})$ melt the system is in the lamellar phase and is described approximately by a single q-mode: $\varphi=\varphi_q exp[iq_0 \cdot r]$. This mean-field free-energy expansion is valid close to the ODT point, but not too close where critical fluctuations become important (8).

The interaction of the melt with confining surfaces is modeled by the following surface integral (9):

$$F_s / kT = \int \sigma(\mathbf{r}_s) \varphi(\mathbf{r}_s) d^2 r_s \qquad (2)$$

At each point $\mathbf{r}_s$ of the surface the interaction is proportional to the polymer density φ, with proportionality constant $\sigma(r)$ (the ``surface parameter''). Thus at points where $\sigma>0$, B monomers are preferred; A monomers are preferred if $\sigma<0$ (10). This surface tension can be tuned, for example, by coating the surface with random copolymers (11). We restrict ourselves for one planar surface located at $z=0$, see Figure 1.

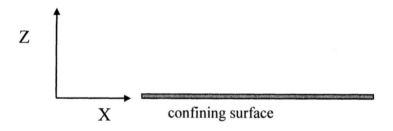

Z

X   confining surface

Figure 1. System of coordinates for a BCP melt confined by one flat surface at $z=0$.

The general spatial dependence of the chemical pattern $\sigma(r_s)$ can be written as (12)

$$\sigma(x,y) = \sum_{\mathbf{q}} \sigma_{\mathbf{q}} e^{i(q_x x + q_y y)} \qquad (3)$$

Here $q=(q_x,q_y)$ is a wavenumber in the plane of the surface. For a melt above the ODT point $(\chi<\chi_c)$, the bulk thermodynamically stable phase is disordered, $\varphi=0$. Hence, for a confined system a linear response theory can be used where the copolymer response function is

$$\varphi(x, y, z) = \sum_{\mathbf{q}} \varphi_{\mathbf{q}}(z) e^{i(q_x x + q_y y)} \tag{4}$$

This density ansatz is put in the free energy $F=F_b+F_s$, which is minimized with respect to each of the functions $\varphi_q(z)$. Neglecting higher order terms (proportional to φ^4), the resulting Euler-Lagrange equation is

$$\left(\tau/h + \left(q_0^2 - q^2\right)^2\right)\varphi_{\mathbf{q}} + 2\left(q_0^2 - q^2\right)\varphi_{\mathbf{q}}'' + \varphi_{\mathbf{q}}'''' = 0 \tag{5}$$

with boundary conditions $(9,12)$

$$\varphi_{\mathbf{q}}''(0) + \left(q_0^2 - q^2\right)\varphi_{\mathbf{q}}(0) = 0$$
$$\sigma_{\mathbf{q}}/h + \left(q_0^2 - q^2\right)\varphi_{\mathbf{q}}'(0) + \varphi_{\mathbf{q}}'''(0) = 0 \tag{6}$$

The general form of $\varphi_q(z)$ is $\varphi_q(z)=Acos(k_1z+\beta)exp(-k_2z)$, which shows decaying oscillations (β is a phase, A is an amplitude). The wavenumbers k_1 and k_2 characterize the periodicity and decay length of the decaying undulations, respectively. They depend on q, q_0, and τ. Once a surface pattern (``picture'') $\sigma(x,y)$ is given, these linear equations are readily solved.

We consider now the copolymer response for several specific chemical patterns. Figure 2 depicts a melt confined by one surface at $z=0$. The pattern in part (a) corresponds to stripe of width d_0 along the y-axis. Inside the stripe the preference is for B monomers (dark) while outside the preference is for A monomers (light). Part (b) shows the melt morphology in the x-z plane. As expected, A monomers (light) are adsorbed to most of the surface, inducing lamellar-like ordering which decays in the z direction. Close to the stripe B monomers are adsorbed, hence creating a disturbance in the overall lamellar layering.

The linear response theory presented by us enables to calculate the copolymer density due to an arbitrary surface pattern (in the x-y plane). Such a surface is shown in Figure 3 (a), where inside the letters ``ACS'' B monomers are preferred and outside the letters the surface is neutral.

There are few points to notice in Figure 3. The morphology is becoming blurred as the distance from the substrate is increased, until it is completely washed out. The morphology for planes separated by a half integer number of bulk lamellae (half integer number of d_0's) are similar, and roughly correspond to an interchange of A and B monomers (dark and light shades). Thirdly, smaller feature sizes disappear more quickly than large features, as the distance from the substrate is increased. This is seen in the corners of the letter ``A'' which are ``smoothed-out'' in (b), (c) and (d) while the overall shape of the letters still persists.

268

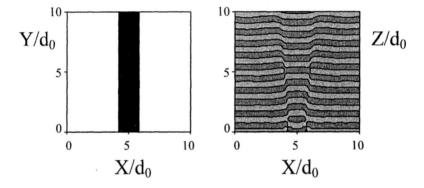

Figure 2. BCP melt confined by one patterned surface. (a) Surface pattern in the x-y plane. Dark stripe prefers B monomers ($\sigma=\frac{1}{2}hq_0^3\varphi_q$). Rest of surface is uniformly preferential to A block ($\sigma=-\frac{1}{2}hq_0^3\varphi_q$). (b) Morphology in the x-z plane: dark shades correspond to B-rich domains while light shades to A-rich domains. The stripe introduces a disturbance to the lamellar ordering of the surface. For $z/d_0>10$ the bulk disordered phase is recovered (not shown). $\chi N=10$.

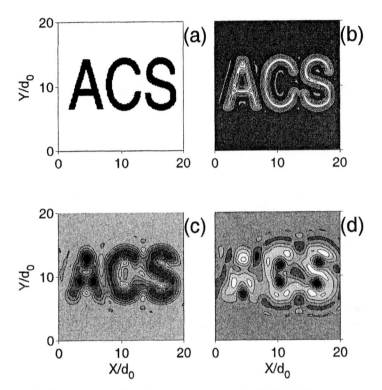

Figure 3. Propagation of surface pattern into the bulk. (a) Substrate pattern in the shape of the letters ``ACS'' (at z=0). The BCP morphology in parallel planes with increasing distance from the substrate is shown in (b), (c) and (d) for z=0.5, 3 and 5 d_0, respectively. $\chi N=9.5$ and $\sigma=hq_0^3\varphi_q$ inside the letters uniformly prefers B monomers and zero outside (neutral).

As we have seen, interactions with confining surface can be useful in achieving a desired BCP morphology. The main experimental parameters allowing control of morphology are temperature, film thickness, polymer molecular weight and block/surface interfacial tension (*13*). In the next section we turn to discuss how electric field can be useful to induce ordering in confined BCP.

BCP in Electric Field: Role of Mobile Dissociated Ions

Electric field offers a fast, easily switchable, means of control over thin-film morphology. Although very effective in the bulk, in a thin film shear is far more problematic, and on the other hand the voltages required are rather modest. For example, a film thickness of 1 μm and a voltage of *100 V* is enough to obtain a high field of *1* V/μm. The effect of electric field on a BCP sample has been extensively studied experimentally (*14,15*) and theoretically (*14,16-18*) in recent years. The main phenomenon examined was orientation of lamellae or cylinders. The orientation results from a torque acting on lamellae/cylinders when these are tilted with respect to the external field E_0. The torque vanishes when the lamellar/cylindrical axis are perpendicular or parallel to E_0, but the mechanically stable configuration is the latter. The reason for torque on a BCP melt stems from the electrostatic free energy penalty associated with dielectric interfaces perpendicular to the field. The strength of the ``dielectric mechanism'' is proportional to $(\varepsilon_A-\varepsilon_B)^2 E_0^2$, where ε_A and ε_B are the dielectric constants of the A and B blocks, respectively. Hence the dielectric mechanism is quadratic in E_0 and vanishes when there is no dielectric contrast, $\varepsilon_A=\varepsilon_B$.

Despite its usefulness and success in predicting experimental results, the dielectric mechanism is rather weak, being proportional to E_0^2. Moreover, in most BCP there are many ion pairs in the material. This is especially true in anionically-prepared copolymers where each chain initially has one BuLi group. At the end of the polymerization process, the chain keeps the Bu and contributes to the surrounding environment one lithium hydroxide (LiOH) group. Moreover, the typical water-rinsing procedure adds even more ions. Some of these groups are dissociated and therefore the melt normally has many mobile charges. We show below that these ions, hitherto ignored in this research field, greatly affect the system behavior. We show that one can use mobile ions to enhance the orientation effect or to induce strong structural changes. Furthermore, a dielectric contrast may not be needed at all, but rather a mobility or a *conductivity contrast* may suffice.

The existence of space charges in electric field introduces Joule heat and dissipation. Hence these systems are intrinsically non-equilibrium ones. We discuss separately the cases of static DC and time-varying AC fields.

BCP in DC Field

Let us demonstrate the effect of dissociated mobile ions in a simple example, a diblock copolymer of polystyrene (PS) polymethylmethacrylate (PMMA) in the bcc phase of spheres (*16*). Domains of the minority phase, PMMA, are sphere-shaped and arranged in a bcc lattice; PS is in the matrix. These isotropic system can not be significantly oriented and therefore the field leads to structural changes. One of the two blocks always has a larger solvation ability than the other. This is usually the high dielectric constant polymer of the two. As mentioned above, dissociated Li and OH ions are present in the melt. Measurements carried out in our groups and by others show that PMMA has superior solvation ability for Li compared to PS. As a result, almost all dissociated ions are located inside the spheres and very few can leak outside to the PS matrix. Our measurements show that a fraction $3 \cdot 10^{-5}$ of all ions are mobile and dissociated, yielding $Q = 10^{-3}$ e per sphere (*19*). Each such sphere has a small dipole $d = 2QR$ consisting of plus charge in one end a negative charge in the other end (R is the sphere radius).

In addition to the contribution from dielectric mechanism there are two new electrostatic interactions: coupling of ions with external field and dipole-dipole interactions between spheres. The first interaction is linear in field E_0 and charge Q. Ions trapped in the spheres (radius $R \approx 10$ nm) optimize electrostatic energy if the spheres become cylinders, because then they can migrate long distances along the field, approximately 1 μm. The second interaction is quadratic in Q and approximately independent of E_0. These interactions destabilize the bcc phase in favor of the cylinder morphology. The critical field for this phase-transition occurs when the electrostatic interaction exactly matches the difference in bulk free energies ΔF (favoring the bcc).

The Leibler Hamiltonian used throughout this chapter enables us to put the numbers in each of these free energy terms. We find that

$$\frac{F_{d-f}}{\Delta F} \cong 1, \qquad \frac{F_{dielec}}{\Delta F} \cong 6 \cdot 10^{-4}, \qquad \frac{F_{d-d}}{\Delta F} \cong 3 \cdot 10^{-5} \qquad (7)$$

In the above F_{d-f} is the dipole-field interaction of ions with the field, F_{dielec} is the free-energy due to dielectric interfaces and F_{d-d} is the dipole-dipole interactions. We have chosen here $N\chi = 12$, $f = 0.37$ and $E_0 = 6$ V/μm. Therefore, the dominant mechanism in the bcc to hexagonal transition is the interaction of ions with the external fields, and for the above parameters it is about 10^3-10^4 times stronger than the dielectric mechanism. The phase-transition with ions is predicted to occur at a field $E_0 \approx 6$ V/μm instead of $E_0 \approx 70$ V/μm when the existence of ions is neglected. Dipole-dipole interactions are negligible. In the next section we generalize these results for time-varying fields.

BCP in AC Field

The physics of BCP in time-varying AC fields is richer than in DC. There are four length scales in the system. The first is d_0, the natural crystal periodicity, $d_0 \approx 50\text{-}200$ nm. The other length scale is $l_{\mathrm{drift}} = \pi e \mu E/\omega$, the drift distance an ion of charge e and mobility μ undergoes in a field E of frequency ω in a time equal to half of the oscillation period. We call the high frequency regime the regime where $l_{\mathrm{drift}} \ll d_0$; the low frequency regime is when $l_{\mathrm{drift}} \gg d_0$. The third length is the wavelength of light, and is given by $l_{\mathrm{light}} = 2\pi c/(\omega\sqrt{\varepsilon})$. In most circumstances l_{light} is larger than the two other lengths, $l_{\mathrm{light}} \gg d_0$, $l_{\mathrm{light}} \gg l_{\mathrm{drift}}$. The last length scale is the system size L. There are three energy scales in the system. One is the energy stored in unit volume of the dielectric material, $U_{\mathrm{dielec}} \equiv \varepsilon E^2$. The second is the Joule heating per unit volume in one field cycle, $U_{\mathrm{Joule}} \equiv 2\pi E^2 \sigma/\omega$. As the frequency of field decreases Joule heating becomes more dominant. The third energy scale is thermal, kT.

In order to calculate forces on the sample one needs to find out the distribution of electric field. In the dynamic Self-Consistent Fields approach, once the electric field is known it is possible to find the contribution to the chemical potential, and then deduce the dynamical behavior from the Laplacian of it (13). Here, however, we take on a different approach where the microscopic force is found directly from the field distribution, which is a function of dielectric constant and mobility distributions. For a weakly-segregated BCP melt, the density variations are small and a linear scheme can be employed, assuming small deviations of field E, dielectric constant ε and mobility μ from their average values E_0, ε_0 and μ_0,

$$\varepsilon(\mathbf{r}) = \varepsilon_0 + \sum_k \varepsilon_k e^{i\mathbf{k}\cdot\mathbf{r}}$$

$$\mu(\mathbf{r}) = \mu_0 + \sum_k \mu_k e^{i\mathbf{k}\cdot\mathbf{r}} \qquad (8)$$

$$\mathbf{E}(\mathbf{r},t) = \mathbf{E}_0 e^{i\omega t} + \sum_k \mathbf{E}_k e^{i(\mathbf{k}\cdot\mathbf{r}+\omega t)}$$

In the low frequency regime Maxwell's equations lead to (17)

$$\mathbf{E}_k = -\frac{8\pi e^2 w \rho_0 \mu_k + \varepsilon_k \omega^2}{8\pi e^2 w \rho_0 \mu_0 + \varepsilon_0 \omega^2} \frac{\mathbf{k}\cdot\mathbf{E}_0}{k^2} \mathbf{k} \qquad (9)$$

In this expression ρ_0 is the average number density of positive and negative ions. Having found the Fourier component of the field we are in a position to calculate the total torque on the sample $L=L_{\text{dielec}}+L_{\text{ions}}$.

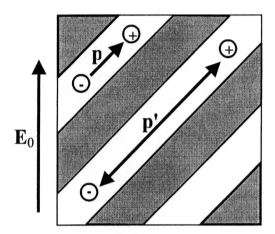

Figure 4. Origin of torque in lamellar BCP in external electric field E_0. The field induces a dipole p due to bound charges. The torque is $E_0 \times p$ and is non-zero if p is not parallel to E_0. In much the same way in AC field free charges (mobile ions) oscillate along stripes, giving rise to a dipole p' and non-zero torque

L_{dielec} is due to torque that external fields exert on dipoles from bound charges, and L_{ions} results from torque of fields and mobile (free) ions, see Figure 4. At a frequency of *50* Hz the ratio between the torque in ion-free BCP and the torque in a realistic ion-containing sample is *(20)*.

$$\frac{L_{ions}}{L_{no\ ions}} = 1 + 16\frac{\sigma_0}{\omega\varepsilon_0} \approx 1.5 \tag{10}$$

$\sigma_0 = e^2\mu_0\rho_0$ is the average sample conductivity. The ratio between the fields is $\sqrt{1.5}$, as $L{\sim}E^2$. This figure emphasizes the importance of mobile ions in orientation experiments in BCP. For frequencies smaller than *50* Hz the mobile ion mechanism becomes increasingly important and it may even dominate the orientation process.

Conclusions

A rich set of pattern formation phenomena exists in diblock copolymers. We demonstrate how bulk morphologies are affected by interacting surfaces or electric fields. The copolymer density as a response to an arbitrary surface chemical pattern is calculated. Each surface q-mode gives rise to a density response mode, which exponentially decays into the bulk. The decay length depends on the temperature and on the q-mode. Small q's (large wavelengths) decays slower than large ones. The distance in which a given surface ``picture'' loses its details is given from our expressions (see, for example, Figure 3).

We next considered the dissipative system of BCPs in electric field, since virtually all polymer show some level of residual conductivity when put under an external potential. Here we are interested in the part of conductivity which is due to mobile dissociated ions. These ions are predicted to cause a phase-transition from bcc phase of spheres into hexagonal phase of cylinders in static DC field. The bcc to hex transition has just been experimentally carried-out in the group of T. Russell in Amherst (21). The surprisingly low fields used, 6-12 V/μm, cannot be explained by dielectric contrast alone, and agree with our theory and confirm our predictions. This kind of deformation should be applicable in many more ordered structures such as some of the triblock copolymer phases.

In AC fields we calculate forces and find that ions lead to a marked increase of the aligning torques acting on the sample. An interesting effect is expected to occur when a gyroid phase is put under the influence of electric field. Due to dielectric and conductive inhomogeneities a phase transition can be realized to the lamellar or hexagonal phases (depending on the exact initial location in the phase-diagram).

A conclusion of this paper is that a *mobility* or *conductivity* contrast rather than a dielectric contrast may be enough to cause alignment in BCPs or other polarizable media. It is our belief that this new and promising mechanism needs to be further explored both theoretically and experimentally.

I would like to thank D. Andelman, J.-B. Fournier, P. G. de Gennes, L. Leibler and F. Tournilhac for illuminating discussions. I thank T. P. Russell for discussions and for communicating his experimental results prior to publication.

References

1. Leibler, L. *Macromolecules* **1980**, *13* 1602.
2. Ohta, T. ; Kawasaki, K. *Macromolecules* **1986**, *19* 2621; Bates, F. S.; Fredrickson, F. S. *Physics Today* **1999**, *52* 32.
3. Seul, S.; Andelman, D. *Science* **1995**, *267* 476.
4. Ouyang Q.; Swinney H. L. *Chaos* **1991**, *1* 411; Lee, K. J.; McCormick, W. D.; Pearson, J. E.; Swinney, H. L. *Nature* **1994**, *369* 215.
5. Ben-Jacob, E. *Nature*, **2002**, *415* 370.
6. Swift, J.; Hohenberg, J. J. *Phys. Rev. A* **1997**, *15* 319.
7. Fredrickson, G. H.; Helfand, E. *J. Chem. Phys.* **1997**, *15* 319.
8. Brazovskii, S. A. *Sov. Phys. JETP* **1975**, *41* 85.
9. Tsori, Y.; Andelman, D. *Europhys. Lett.* **2001**, *53* 722.
10. Pickett, G. T.; Balazs, A. C. *Macromolecules* **1997**, *30* 3097; Menelle, A.; Russell, T. P.; Anastasiadis, S. H.; Satija, S. K.; Majkrzak, C. F. *Phys. Rev. Lett.* **1992**, *68* 67.
11. Kellogg, G. J.; Walton, D. G.; Mayes, A. M.; Lambooy, P.; Russell, T. P.; Gallagher, P. D.; Satija, S. K. *Phys. Rev. Lett.* **1996**, *76* 2503.
12. Petera, D.; Muthukumar, M. *J. Chem. Phys.* **1998**, *109* 5101; Tsori, Y.; Andelman, D. *Macromolecules* **2001**, *34* 2719.
13. Knoll, A; Horvat, A; Lyakhova, K. S.; Krausch, G; Sevink, G. J. A.; Zvelindovsky, A. V.; Magerle, R. *Phys. Rev. Lett.* **2002**, *89* 35501.
14. Amundson, K.; Helfand, E.; Quan, X.; Smith, S. D. *Macromolecules* **1993**, *26* 2698; Amundson, K.; Helfand, E.; Quan, X.; Hudson, S. D.; Smith, S. D. *Macromolecules* **1994**, *27* 6559.
15. Mansky, P.; DeRouchey, J.; Russell, T. P.; Mays, J.; Pitsikalis, M.; Morkved, M.; Jaeger, H. *Macromolecules* **1998**, *31* 4399; Böker, A.; Knoll, A.; Elbs, H.; Abetz, V.; Müller, A. H. E.; Krausch, G. *Macromolecules* **2002**, *35* 1319.
16. Pereira, G. G.; Williams, D. R. M. *Macromolecules* **1999**, *32* 8115.
17. Ashok, B.; Muthukumar, M.; Russell, T. P. *J. Chem. Phys.* **2001**, *115* 1559.
18. Tsori, Y.; Andelman, D. *Macromolecules* **2002**, *35* 5161.
19. Tsori, Y.; Tournilhac, F.; Andelman, D. ; Leibler, L. accepted to *Phys. Rev. Lett.* **2002**.
20. Tsori, Y.; Tournilhac, F.; Leibler, L. submitted to *Macromolecules* **2002**.
21. Russell, T. P. Private communication.

Chapter 22

Phase Separation of Polymer Blends Driven by Temporally and Spatially Periodic Forcing

Qui Tran-Cong-Miyata, Shinsuke Nishigami, Shinsuke Yoshida, Tetsuo Ito, Kazuhiro Ejiri, and Tomohisa Norisuye

Department of Polymer Science and Engineering, Kyoto Institute of Technology, Matsugasaki, Kyoto 606–8585, Japan

Phase separation of polystyrene/poly(vinyl methyl ether) blends was induced under spatially and temporally periodic forcing conditions by taking advantage of either photodimerization of anthracene or photoisomerization of *trans*-stilbene chemically labeled on polystyrene chains. Significant mode-selection processes driven by these reactions were experimentally observed and analyzed for both cases. These experimental results reveal a potential method of morphology control using periodic forcing conditions.

Introduction

Phase separation of polymer mixtures occurs through a process in which the fluid mixture evolves through a series of states between the homogeneous fluid mixture and a macroscopically phase separated, stratified state. In general, when a mixture becomes unstable under external perturbations such as variation of temperature or pressure, arise composition fluctuations of various wavelengths. The initial scale of the fluctuations near critical composition is comparable to the interfacial width between the phases at equilibrium and these fluctuations grow with time through diffusional process under the constraint of volume conservation. The growing morphology is dominated by the most unstable fastest growing modes which are sensitive to external perturbation (1). For most cases, phase separation of polymer blends terminates at random, isotropic morphology controlled by the free-energy minimum of the systems .

The above-mentioned scenario is greatly modified by the presence of chemical reactions. It has been shown recently by numerical as well as analytical calculations that chemical reaction can be used as a driver for these unstable modes (2-7)., suggesting a novel method for morphology control of polymer mixtures. Experimentally, we have demonstrated that not only the characteristic length scales (8), but also the spatial symmetry of the morphology can be manipulated by taking advantages of photochemical reactions (9-10).

Chemical reactions in bulk polymers can generate an internal stress in these materials in the course of reaction (11, 12). This stress field can couple to the concentration fluctuations associated with phase separation, resulting in complex phase separation kinetics and morphology. The complicated effects of elastic stress on phase separation kinetics have been reported previously in the case of photoisomerization (8). Most recently, the stress field developing inside photo-cross-linked polymer blends has been observed in situ and analyzed by using Mach-Zehnder interferrometry (13). The effects of elastic stress field were also examined in details for mixtures undergoing polymerization in the presence of liquid crystals (14). Since the elastic stress can actually generate long-range effects on the phase separation, it is expected that the stress relaxation plays an important role in the mode selection process of phase separation and therefore in the resulting morphology.

We present here a study on phase separation of polymer mixtures forced with temporally and spatially periodic irradiation. The main purpose is to elucidate the contribution of the elastic stress associated with the changes in polymer structure generated by chemical reactions. In the temporal modulation experiments, phase separation was induced by periodic irradiation using ultraviolet (uv) light chopped with frequency varying between 1/200 to 100 Hz.

On the other hand, in the spatial modulation experiments, the mixtures were irradiated with continuous uv light through photomasks with different spacing ranging from 8 mm to 10 μm. This spatial modulation generates periodic regions with different elastic modulus in the irradiated blend corresponding to the dark and bright areas of the blend. Under these conditions, we also have spatial confinement effects imposed on the phase separation by the limited spacing of the photomasks. In this chapter, the experimental results obtained with both temporally and spatially forcing are summarized and discussed in terms of mode-selection driven by chemical reactions.

General Mechanism of Phase Separation Induced by Chemical Reactions

In general, introducing chemical reactions, such as chemical crosslinking between the polymer chains, into a polymer mixture destabilizes these mixtures, leading to phase separation. The most well-known example is the phase separation accompanying with polymerization and/or the network formation during the synthesis of interpenetrating polymer networks (IPN) (15). For photochemical reactions, it has been demonstrated that photo-cross-linking (16, 17) as well as photoisomerization (18, 19) can be used to trigger and control phase separation of polymer blends. The general mechanism is the increase in the Gibbs free-energy (G) of the mixture triggered by the reaction. Compared to mixtures of small molecules, a large change in G can be easily generated by chemical modification of polymer. This normally unfavorable change mainly comes from the decrease in entropy of the mixtures during polymerization or cross-link (20). For the case of photoisomerization, an increase in the Gibbs free-energy can be created by the changes in segmental volumes and/or in the enthalpic interactions between polymer segments associated with the trans→ cis isomerization (19). By coupling these photochemical reactions to phase separation, light can be efficiently utilized as a tool to generate as well as manipulate morphology of polymer blends.

Samples Synthesis and Characterization

Samples used in this work are the mixtures of poly (vinyl methyl ether) (PVME, $M_w = 1.0 \times 10^5$, $M_w/M_n = 2.5$) and polystyrene (PS) derivatives. To be able to induce phase separation by light, the PS component was chemically labeled with photoreactive groups, either with anthracene (A) or trans-stilbene (S). The synthesis procedure of these polymers was published elsewhere (8, 16).

The characteristics of these photoreactive polymers are as follows: PSA (M_w= 2.2 x 10^5, M_w/M_n = 1.7; label content 1.8 mole%), PSS (M_w= 2.8 x 10^5, M_w/M_n = 1.8; label content 4.3 mole%). Similar to their PS/PVME "parent" mixture, all these photosensitive polymer blends exhibit a lower critical solution temperature (LCST). Specifically, they undergo phase separation upon increasing temperature. Their phase behavior is provided in the subsequent sections described below.

Temporally Periodic Forcing of Phase Separation

Experiments on temporally forcing of phase separation were performed using mixtures of poly(vinyl methyl ether) (PVME) and anthracene-labeled polystyrene (PSA). The phase diagram of this PSA/PVME blend is illustrated in Figure 1 together with the composition dependence of its glass transition

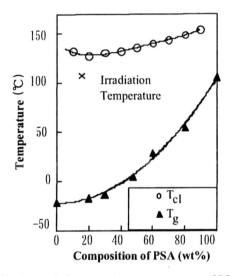

Figure 1. Cloud points and glass transition temperatures of PSA/PVME blends.

temperatures (T_g). Irradiation experiments were carried out in the one-phase region above T_g of the blend. UV light from a Hg-Xe lamp (350W, Moritex, Japan) was first passed through appropriate optical filters to extract the 365 nm component and was subsequently impinged periodically onto the blends by using an optical chopper (Stanford Research Systems, Inc.) operating at frequencies (f) ranging from 1/120 to 200 Hz. The blend was kept in a heating block at a constant temperature that was regulated with a precision of $\pm 0.5°C$. Due to the specific structure of the chopper, the *on* and *off* periods of irradiation have equal time intervals. This leads to the consequence that the total number of photons received by the blend under these periodic irradiation conditions is exactly one half of the case of continuous irradiation. The light intensity used in this work was fixed at 3.0 mW/cm², unless otherwise noted.

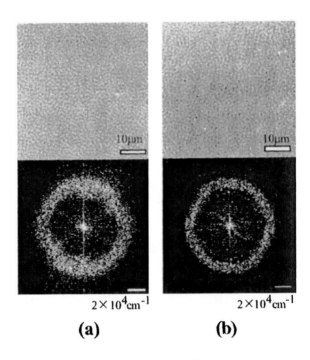

$2 \times 10^4 \text{cm}^{-1}$ $2 \times 10^4 \text{cm}^{-1}$

(a)   **(b)**

Figure 2. Morphology and the corresponding 2D-Fourier intensity distribution (power spectra) of a PSA/PVME(20/80) blend obtained at 100°C by: (a), continuous irradiation in 75 min; (b), periodic irradiation in 150 min with f=50 Hz.

Figure 2-a shows the morphology of a PSA/PVME (20/80) blend observed by using phase-contrast optical microscopy (Nikon, XF-NTF-21) after 75 min of continuous irradiation at 100°C (i.e. 20 °C below the cloud point). The corresponding power spectra obtained by 2-dimensional fast Fourier transform (2D-FFT) of the optical micrograph are illustrated in the same figure. The presence of the so-called "spinodal" ring suggests that the spinodal decomposition induced by irradiation was frozen by the photodimerization of anthracene moieties labeled on polystyrene chains. On the other hand, the morphology and the corresponding 2D-FFT power spectra of the same PSA/PVME (20/80) blend irradiated periodically with the frequency 50 Hz under the same conditions are illustrated in Figures 2b. We observe that the "width" of the spinodal ring obtained under periodic irradiation is much narrower compared to the data of the same blend receiving the same amount of photons from the same light source. This result indicates that the wavelength distribution of various periodic structures inside the spinodal structure of the blend can be modified by the irradiation frequency. In order to quantitatively characterize the distribution of these characteristic length scales, the Fourier intensity distribution of the optical image was fitted to the following modified Gaussian distribution :

$$I(q) = \frac{a}{\sqrt{2\pi\sigma^2}} \exp[\frac{(q - q_{max})^2}{2\sigma^2}] + b q^{-c} + d \qquad (1)$$

where

$I(q)$: the Fourier intensity distribution of the optical image obtained from 2D-FFT.

σ: the distribution of the characteristic length scales of the morphology.

a: a constant characterizing the contrast of the image.

q_{max} : the frequency corresponding to the maximal Fourier intensity.

b, c, d : the constants characterizing the decay of the Fourier intensity with respect to the wavenumber q.

It was found that the above equation provides a very good fit for the experimental data. Also, there exists a particular irradiation frequency where the width of the morphological length scales distribution passes through a minimum. As an example for this modulation effect, the half-width at half-maximum (hwhm) of the 2D power spectra was calculated from the dispersion parameter σ in Eq. (1). The result obtained for a PSA/PVME (15/85) blend at 110°C is shown in Figure 3 for several modulation frequencies ranging from 0 Hz (corresponding to the case of continuous irradiation) to 100 Hz. On the other hand, the dependence of these characteristic length scales on the modulation frequency is not significant, suggesting that the morphology of the irradiated blend was frozen by the photo-cross-linking reactions under these experimental conditions. The magnitude of the Fourier intensity which reflects the contrast of the morphology, i.e. the amplitude of these periodic structures, varies in accordance with the behavior of the length scale distribution. Namely, the contrast of the morphology becomes strongest, i.e. log C(q) goes through a maximum, at the irradiation frequency where the width σ passes a minimum.

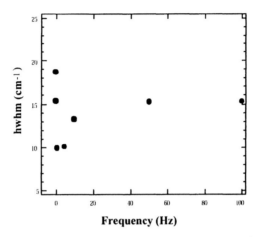

Frequency (Hz)

Figure 3. Irradiation dependence of the half-width at half-maximum (hwhm) calculated from the parameter σ in Eq. (1) for a PSA/PVME (15/85) blend irradiated at 110°C.

These experimental results suggest the existence of a mode-selection process in the irradiated blend driven by external modulation frequency. From the kinetic viewpoint, by cross-linking the mixture with a modulating frequency f, all the unstable modes with the growth rates smaller than f would be arrested by the cross-linking and are then stabilized. On the other hand, for those modes which develop much faster than the modulation fequency, the reaction would be ineffective in selecting a particular structure with a finite range of wavelengths. We suggest that this narrowing process originates from the relaxation of the local stress generated by the inhomogeneous cross-linking kinetics in the bulk state of polymeric systems *(21)*. By introducing an intermittency (the *off-period*) into the irradiation process, the extent of the relaxation of this elastic stress can be controlled by the irradiation frequency. These experimental results suggest not only a novel method of controlling the morphological length scales distribution in multiphase polymers, but also reveal the potential of designing polymer materials for photonic applications.

Spatialy Periodic Forcing of Phase Separation

During phase separation process, the morphology of polymer blends develops from unstable concentration fluctuations. The creation and annihilation of these fluctuations during this stage form a basis for the mode-selection governing the morphology. For non-reacting binary mixtures, the length scales of the morphology emerging at the beginning of phase separation are determined by the Cahn-Hilliard critical wavelengths *(22)*. For polymer mixtures, these initial length scales can be obtained by linear stability analysis of kinetic equations using appropriate Gibbs free-energy models *(23, 24)*. It is expected that phase separation kinetics and the corresponding morphology will be greatly modified if the mixture is directionally confined into a certain length scale close to these Cahn-Hilliard critical wavelengths. Recently, the effects of spatial confinement on phase separation of polymer blends have been investigated by using chemically patterned substrates and the pattern scale selection phenomena were clearly demonstrated in these studies (25-27). Here, instead of chemically patterning, we use uv light to induce phase separation of polymer blends in a specific region at the micrometer scale. For this purpose, miscible blends of stilbene-labeled polystyrene (PSS) and poly(vinyl methyl ether) (PVME) were forced to undergo phase separation by irradiation through photomasks with different spacings. Specifically, PSS/PVME blends were irradiated with 325 nm uv light from a He-Cd cw laser (Nihon Lasers In c.) through photomasks with various spacing in the range 100 - 20 μm. The largest spacing of the photomasks is 8 mm corresponding to the case of "free-from-forcing". The sample was sandwiched between two pieces of cover glass with a 30 μm aluminum spacer used to adjust the sample thickness. All the experimental and analysis procedure are similar to those described in the previous session. Particularly, two-dimensional laser light scattering using a He-Ne laser was also monitored to confirm the morphology obtained by optical microscopy in real space.

Similar to its "parent" mixture, PSS/PVME blends exhibit a lower critical solution temperature. The phase diagram and the glass transition temperature of PSS/PVME blends are illustrated in Figure 4. Irradiation experiments were performed at 95°C in the miscible (one-phase) region above the glass transition temperature of the blend. Shown in Figures 5 are the morphology and its corresponding 2D-FFT power spectra obtained for a PSS/ PVME (20/80) blend irradiated using 3 kinds of photomasks with the spacings (Λ) = 8mm, 50 μm and 20 μm. There clearly exists a transition from isotropic to anisotropic morphology as the spacing Λ is reduced below 50 μm. When Λ is very large (8 mm) compared to the intrinsic characteristic length scales of the phase separated stru cture (ca.

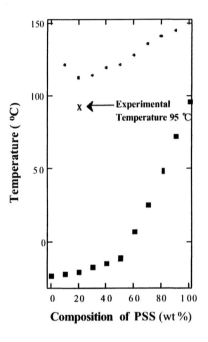

Composition of PSS (wt %)

Figure 4. Cloud points and glass transition temperatures of PSS/PVME blends.

few microns), the resulting morphology is isotropic as revealed by the so-called "spinodal ring" illustrated in Fig. 5-d. As the spacing becomes narrower, the 2D power spectra gradually turn into arc-shape and eventually exhibit spots aligning along the x-direction when Λ reaches 20 μm. It should be noted that no phase separation was observed with $\Lambda = 4.6$ μm when an interference pattern generated by crossing two He -Cd laser beams was used for irradiation. Furthermore, morphology obtained by irradiation with $\Lambda = 20$ μm exhibits multiple length scales as shown in Fig. 5-c where there exist two types of periodic structures. One (*macro-stripe*) has the period equivalent to the spacing of the photomask and the other (*micro-stripe*) has much shorter period . In order to confirm the presence of this micro-stripe structure, we utilize a 2D light scattering photometer to monitor the diffraction patterns from the irradiated polymer blend. Shown in Figure 6 is the 2D diffraction pattern obtained for a PSS/PVME (20/80) blend irradiated through a 20 μm photomask. Two specific features were observed with this particular diffraction pattern. One is the diffraction pattern from the macro-lattice characterized by multiple strong spots detectable upto 41st order along the x-direction. This diffraction pattern also exhibits several minima which are approximately located at the n. 7th orders where n is an integer. These characteristics suggest that the macro-grating generated in the irradiated blend does not have equal (1:1) volume fraction, but it is approximately 3:4 *(28)*. The second feature is the existence of the strong diffraction around 3.61 μm^{-1} which is located approximately at the position of the 23rd order diffraction of the macro-lattice. This secondary maximum corresponds to a Bragg spacing of ca. 1.7 μm which is in good agreement with the characteristic length scale of the micro-stripe structure observed by phase-contrast microscopy shown in Fig. 5-c. It should be noted that the origin of these two kinds of pattern scales is completely different. The macro-stripe pattern is generated by the spatial modulation of refractive index of the *trans*-and *cis*-isomers of stilbene in the blend. This refractive- index distribution comes from the spatial modulation of the concentrations of *trans*- (in the bright regions) and *cis*-stilbene (in the dark regions) which only exists after irradiation. On the other hand, the micro-stripe shown in Fig. 5-c is probably caused by the restriction imposed on the phase separation of the blend along the x-axis by the photomasks. Namely, as the length scale under which phase separation is induced by irradiation, approaches the critical Cahn-Hilliard wavelengths, unstable modes of concentration fluctuations might be partially stabilized. This is because the unstable fluctuations cannot grow beyond the limit imposed by the spacing of photomask along the x-direction, whereas there is no restriction along the y-direction. This asymmetrical destabilization of concentration fluctuations in the reacting blend results in the anisotropic morphology shown in Figs. 5. It is worth noting that the morphological anisotropy described above is a consequence of phase separation and is not originated from the so-called "relief grating" reported recently for homopolymers irradiated with interference patterns of uv laser *(29, 30)*. To confirm the difference between the mechanism of the structure formation observed in our study and of the relief grating obtained in laser patterning experiments, a PSS/ PVME (20/80) blend was first irradiated by using a He-Cd laser through a photomask with 20 μm spacing. Subsequently, the blend was annealed under vacuum in the dark for 2 hrs at 110°C which is located at 130°C above the glass transition temperature (Tg) of the blend. It was found that the micro-stripe structure seen in Figs. 5 did not disappear, but instead coarsened with an increase in contrast. Taking into account that the relief grating formed in homopolymers irradiated with laser at ambient temperature and subsequently annihilated upon heating at temperatures above Tg of the irradiated polymer, the above experimental results suggest that the pattern with shorter period observed in Figs. 5 is the consequence of phase separation induced by light at the length scale restricted to μm. More experiments such as changing the light intensity, the irradiation temperatures and the reaction kinetics (cross-link vs. non-cross-

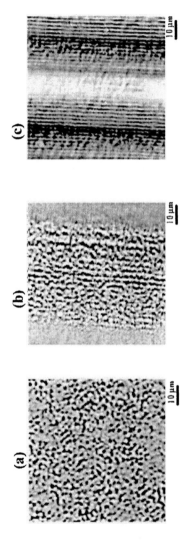

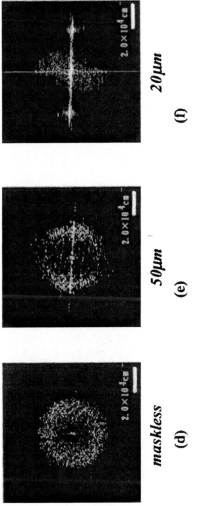

maskless

(d)

50μm

(e)

20μm

(f)

Figure 5. Phase-contrast micrograph and the corresponding 2D-FFT power spectra of a PSS/PVME (20/80) obtained at 95°C by irradiation using photomasks with various spacing. The light intensity is $30mW/cm^2$.

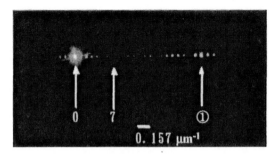

Figure 6. Two-dimensional light scattering pattern of a PSS/PVME(20/80) blend with the morphology shown in Fig. 5-c. The light source is a He-Ne laser.

link) are currently in progress to understand the mechanism of phase separation under spatial confinement.

Concluding Remarks

We have demonstrated the effects of spatial as well as temporal modulation on the phase separation of photo-reactive polymer blends. The principal findings are as follows:

1) In the phase separation induced by temporal modulation, it was found that there exists a particular range of irradiation frequency under which the phase morphology becomes quite regular. This indicates that the characteristic length scale distribution of the morphology can be controlled by using external modulation. These results also suggest the important contribution of the stress relaxation to the morphology induced by chemical reactions.

2) Phase separation of polymer blends starts being influenced by the spatial confinement when the scale (Λ) of phase separation becomes smaller than 50 μm. For $\Lambda = 20$ μm, the morphology is a stripe structure oriented perpendicularly to the direction of spatial restriction. As the spacing goes beyond 10 μm, no phase separation was observed by optical microscope as well as by light scattering. These experimental results suggest the existence of a mode selection process driven by the spatial restriction on phase separation of polymer blends.

Acknowledgment

This work is financially supported by the Ministry of Education, Culture, Sports, Science and Technology (MONKASHO) Japan (to Q. T.-C.-M) through

289

the Grant-in-Aid No. 13031054 on the Priority-Area-Research *"Dynamic Control of Strongly Correlated Soft Materials"*. The technical advise of Professor Okimichi Yano (Department of Polymer Science and Engineering, KIT) and very helpful discussion with Professor Yushu Matsushita (Department of Applied Chemistry, Nagoya University) are gratefully appreciated. Particularly, we would like to thank the Iwatani Naoji Foundation (Tokyo, Japan) for the generous support for this Symposium.

References

1. For example, see Onuki, A. *Phase Transition Dynamics*; Cambridge University Press: Cambridge, U.K., 2002.
2. Glotzer, S.C. ; Di Marzio, E.A.; Muthukumar, M. *Phys. Rev. Lett.* **1995**, *74*, 2034.
3. Verdasca, J.; Borckmans, P.; Dewel, G. *Phys. Rev. E* **1995**, *52*, 4616.
4. Carati, D.; Lefever, R. *Phys. Rev. E* **1997**, *56*, 3127.
5. Motoyama, M; Ohta, T. *J. Phys. Soc. Jpn.* **1997**, *66*, 2715.
6. Hildebrand, M.; Mikhailov, A.S.; Ertl, G. *Phys. Rev. E* **1998**, *58*, 5483.
7. Reigada, R.; Sagues, F.; Mikhailov, A.S. *Phys. Rev. Lett.* **2002**, *89*, 038301.
8. Ohta, T.; Urakawa, O.; Tran-Cong, Q. *Macromolecules* **1998**, *31*, 6845.
9. Kataoka, K.; Urakawa, O.; Nishioka, H.; Tran-Cong, Q. *Macromolecules* **1998**, *31*, 6845.
10. Nishioka, H.; Kida, K.; Yano, O.; Tran-Cong, Q. *Macromolecules* **2000**, *33*, 4301.
11. Smets, G. *Adv. Polym. Sci.* **1983**, *50*, 17.
12. Kataoka, K.; Harada, A.; Tamai, T.; Tran-Cong, Q. *J. Polym. Sci. Polym. Phys.* **1998**, *36*, 455.
13. Inoue, K.; Komatsu, S.; Norisuye, T.; Tran-Cong-Miyata, Q. to be published.
14. Nakazawa, H.; Fujinami, S.; Motoyama, M.; Ohta, T.; Araki, T.; Tanaka, H.; Fujisawa, T.; Nakada, H.; Hayashi, M.; Aizawa, M. *Comp. Theor. Polym. Sci.* **2001**, *11*, 445.
15. *Interpenetrating Polymer Networks* ; Klempner, D.; Sperling, L.H.;Utracki, L.A. Eds.; Advances in Chemistry Ser. No. 239, American Chemical Society: Washington, DC, 1994.
16. Harada, A.; Tran-Cong, Q. *Macromolecules* **1997**, *30*, 1643.
17. Turner, J.S.; Cheng, Y. -L. *Macromolecules* **2003**, *36*, 1962.
18. Irie, M.; Iga, R. *Makromol. Chem. Rapid Commun.* **1986**, *7*, 751.
19. Urakawa, O., Yano, O.; Tran-Cong, Q.; Nakatani, A.I.; Han, C.C. *Macromolecules* **1998**, *31*, 7962.
20. Binder K.; Frisch, H. L. *J. Chem. Phys.* **1984**, *81*, 2126.
21. Kataoka, K.; Urakawa, O.; Nishioka, H.; Tran-Cong, Q. *J. Polym. Sci. Polym. Phys.* **1998**, *36*, 455.
22. Cahn, J.W. *Acta Metall.* **1961**, *9*, 795; ibid. **1962**, *10*, 179.

23. Flory, P.J. *Principles of Polymer Chemistry;* Cornell University Press: Ithaca, New York, 1953; Chapters 12-13.
24. de Gennes, P.G. *J. Chem. Phys.* **1980**, *72*, 4756.
25. Ermi, B.D.;Nisato, G.; Douglas, J.F.; Rogers, J.A.; Karim, A. *Phys. Rev. Lett.* **1998**, *81*, 3900.
26. Nisato, G.; Ermi, B.D.; Douglas, J.F.; Karim A. *Macromolecules* **1999**, *32*, 2356.
27. Sehgal, A.; Ferreiro, V.; Douglas, J.F.; Amis, E.J.; Karim, A. *Langmuir* **2002**, *18*, 7041.
28. Hosemann, R.; Bagchi, S.N. *Direct Analysis of Diffraction by Matter;* North-Holland Publishing Co.: Amsterdam, 1962.
29. Rochon, P.; Batalla, E.; Natansohn, A. *Appl.Phys. Lett.* **1995**, *66*, 136.
30. Kim, D.Y.; Li, L.; Kumar, J.; Tripathy, S.K. *Appl.Phys. Lett.* **1995**, *66*, 1166.

Oscillatory Systems

Chapter 23

Perturbation of the Oscillatory Belousov–Zhabotinsky Reaction with Polyethylene Glycol

Renato Lombardo[1], Carmelo Sbriziolo[1], Maria L. Turco Liveri[1,*], Krisztina Pelle[2], Maria Wittmann[2], and Zoltan Noszticzius[2]

[1]Dipartimento di Chimica Fisica "F. Accascina", Università di Palermo, Viale delle Scienze–Parco D'Orleans II, 90128 Palermo, Italy
[2]Center for Complex and Nonlinear Systems and the Department of Chemical Physics, Budapest University of Technology and Economics, H–1521 Budapest, Hungary

The effects of the water soluble non-ionic polymers poly(ethylene glycol) (PEG-400, PEG-900 and PEG-2000) and poly(ethylene glycol dimethylether) (MPEG-2000) on the dynamics of the cerium ion catalyzed oscillatory Belousov-Zhabotinsky (BZ) reaction were investigated in a stirred batch reactor by measuring CO_2 evolution rate and Ce(IV) absorbance changes. Addition of increasing amounts of additives strongly influences all the oscillatory parameters (induction period, time period and amplitude) to an extent that depends on the type of the additive used. The polymer perturbation effects are compared with those previously obtained with methanol and ethylene glycol (1). All these experimental observations can be understood qualitatively by model calculations regarding i) the reactions of the alcoholic endgroups which generate the autocatalytic bromous acid and ii) the reactions of the polymer backbone which produce hypobromous and bromomalonic acids.

Introduction

It is well known that the Belousov-Zhabotinsky (BZ) reaction can initiate free radical polymerisation (2) while it is less known that polymers can also affect the dynamics of the BZ reaction. Recently, we have performed preliminary experiments perturbing the BZ reaction with two different water-soluble non-ionic polymers containing alcoholic end-groups, namely polypropylene glycol and polyethylene glycol (PEG) (3). It was realized that the Belousov-Zhabotinsky reaction responded to the perturbation in an unexpected way. Thus, a systematic study was undertaken to inquire whether the perturbation effect can be attributed exclusively to PEG reactive endgroups (here: primary alcoholic groups) or the chemical nature of polymeric backbone plays also a relevant role.

As a first step (1), in order to study the effects of the alcoholic end-group separately from those due to the polymer nature we have perturbed the BZ reaction with the low molecular weight (LMW) alcohols: ethylene glycol (EG), the monomer of PEG, and methanol (MeOH), a more simple primary alcohol. It has been found that on a qualitative scale the two alcohols bring about the same perturbation effects on the dynamics of the BZ system. In particular, presence of alcohol in the BZ mixture increases the induction period while both the oscillation period and the amplitude of oscillations are decreased. At a given concentration both EG and MeOH prevent the system from oscillating. Comparison of the experimental results with model calculations applying a latest model (4) of the BZ reaction, the Marburg-Budapest-Missoula (MBM) mechanism extended with the perturbing reactions, showed a good agreement on a qualitative scale (alcoholic perturbations increase the induction period and the frequency of the oscillations and decrease the amplitude) but disagreements were found on a quantitative level. Because the mechanism of the alcoholic perturbation, especially in the case of methanol, is mostly clarified (5), we have suggested that probably some minor modifications of the MBM mechanism should be made in the future.

The present work is the next step where we have undertaken a systematic study of the perturbed BZ system with an alcoholic and a nonalcoholic polymer, namely polyethylene glycol with different molecular weights and polyethyleneglycol dimethylether (MPEG), respectively. The dynamical behaviour of the BZ system has been monitored by means of CO_2 evolution rate and spectrophotometric measurements. Moreover, for the sake of comparison we have also performed spectrophotometric measurements in the presence of ethylene glycol.

As the reaction dynamics responds to the polymer perturbations rather sensitively simulating these experiments can help to test both the MBM mechanism and the proposals for the mechanism of the perturbation.

The versatile polyethylene glycol has been chosen since we think that it gives a better chance to compare the polymer - monomer and the polymer backbone - endgroup effects. In fact, the ratio of the reactive endgroups to the polymer backbone can be easily varied by changing the polymer molecular weight. In addition, the methoxylated polymer allows to eliminate the perturbation effects due the –OH terminal groups.

Experimental Methods

Potassium bromate, $Ce(SO_4)_2$ $4H_2O$, malonic acid (MA), ethylene glycol (EG), polyethylene glycol (PEG) with molecular weight 400, 900 and 2000 g mol^{-1} (PEG-400, PEG-900 and PEG-2000, respectively), polyethyleneglycol dimethylether with molecular weight 2000 g mol^{-1} (MPEG-2000) and sulfuric acid were of commercial analytical quality (Fluka) and used without further purification. Deionized water from reverse osmosis (Elga, model Option 3), having a resistivity higher than 1 MΩ·cm, was used to prepare all solutions. Stock solutions of sulfuric acid were standardized by acid-base titration.

The oscillating mixtures for kinetic runs were obtained by mixing freshly prepared aqueous stock solutions of $KBrO_3$, additive (at the desired concentration) with Ce(IV) in 2.00 mol dm^{-3} sulfuric acid and MA in 2.00 mol dm^{-3} sulfuric acid.

The following concentrations of reactants in the final solutions were used: [MA] = 0.1 mol·dm^{-3}, [KBrO$_3$] = 0.03 mol·dm^{-3}, [Ce(IV)] = 4·10^{-4} mol·dm^{-3}, [H$_2$SO$_4$] = 1 mol·dm^{-3}, 1.87·10^{-4} mol·dm^{-3} $\leq$ [additive] $\leq$ 3.75·10^{-2} mol·dm^{-3}.

The current study utilizes two complementary techniques to monitor the behaviour of the BZ system in the absence and presence of additives: CO_2 evolution rate measurements (6), which reflects the rate of the overall reaction, and UV-vis spectrophotometric measurements.

CO$_2$ evolution measurements. CO$_2$ evolved during the reaction has been removed from the reactor by a stream of nitrogen carrier gas bubbled through the solution that provided also the stirring. The CO$_2$, mixed with hydrogen, was converted to methane on a nickel catalyst and measured by a flame ionisation detector (FID) giving a CO$_2$ evolution rate vs. time plot. The temperature of all the experiments was kept at 20.0 ± 0.5°C.

UV-vis spectrophotometric measurements. The changes in the Ce(IV) absorbance at 350 nm have been recorded with a computer-controlled Beckman model DU-640 spectrophotometer, equipped with thermostated compartments for 1.00 cm cuvettes and an appropriate magnetic stirring apparatus. The temperature of all the experiments was kept at 20.0 ± 0.1 °C.

Results and Discussion

Figure 1 shows typical spectrophotometric and CO_2 evolution measurements of an unperturbed and a polymer-perturbed BZ system. The experimental results obtained with the two different techniques are in fair agreement.

The effect of the polymer perturbation resembles to that of the low molecular weight alcohols. Distinct differences can be also noticed when comparing the effect of the monomer and the polymer. For example in our previous experiments with LMW alcohols (1) it was found that the alcoholic perturbation increases the induction period (IP) and decreases both the oscillation period (τ) and the amplitude (A) in a monotonic way. In the case of polymer perturbants the situation is more complex. To illustrate the deviations between the various perturbants Figures 2 shows that the IP

1. for PEG-400, is almost constant up to a concentration around $9.37 \cdot 10^{-3}$ mol·dm^{-3} and then significantly increases
2. for PEG-900 and PEG-2000, decreases up to a concentration of around $9.37 \cdot 10^{-3}$ mol·dm^{-3} and then drastically increases, although the decrease is less pronounced in the case of PEG-900
3. always decreases on increasing MPEG-2000 concentration
4. always increases on increasing EG concentration.

Polymers are less effective in lengthening the induction period than the LMW alcohols. However, for all alcoholic polymers used, beyond a critical concentration value the oscillations were completely suppressed like in the case of LMW alcohols (1).

As to the influence of polymers on the other oscillatory parameters it has been found that on increasing the additive concentration

1. for all polymers used the oscillation period decreases (Figure 3). Perusal of Figure 3 suggests that the extent of the decrease is more pronounced for the additives having –OH terminal groups
2. both for PEGs and EG the amplitude decreases while it remains almost unchanged in the case of MPEG-2000.

These findings suggest that there should be certain reaction(s) of the polymer which does not take place with the monomer in the BZ reaction. In that case PEG perturbs the BZ system not only by its reactive side groups, *i.e.* the

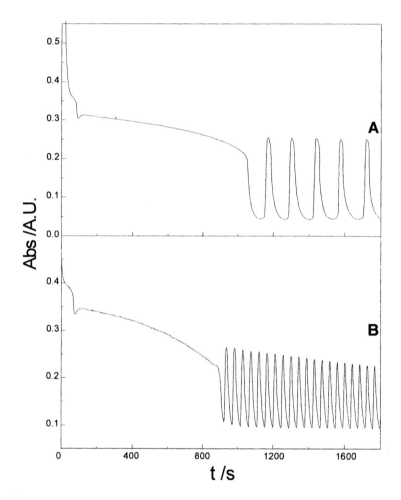

Figure 1. Time dependence of the Ce(IV) absorbance and CO_2 evolution rate in the absence (A and C) and presence (B and D) of 0.015 mol·dm^{-3} PEG-2000. Initial concentrations of the BZ system: $[MA]_0$ = 0.1 mol·dm^{-3}, $[KBrO_3]_0$ = 0.03 mol·dm^{-3}, $[Ce(IV)]_0$ = 4·10^{-4} mol·dm^{-3}, $[H_2SO_4]$ = 1 mol·dm^{-3}.

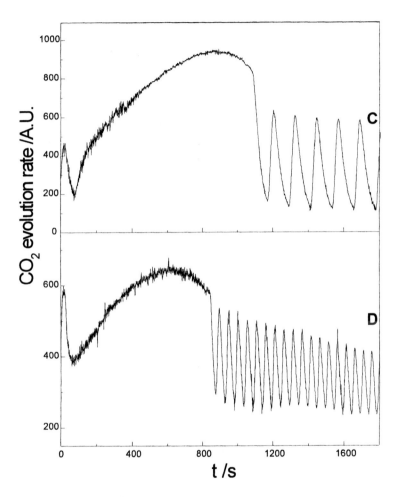

Figure 1. *Continued.*

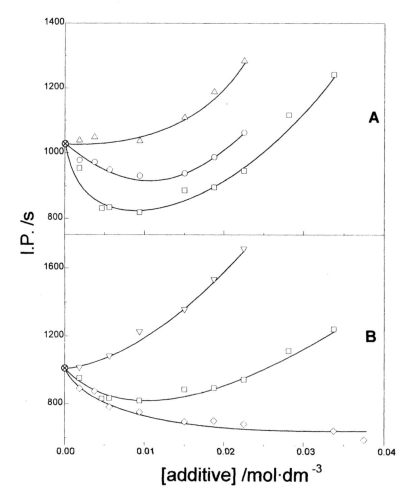

Figure 2. Plot of the length of the induction period (IP) as a function of the additive concentration for the perturbed BZ system with (∇ and ▼) EG, (△ and ▲) PEG-400, (○) PEG-900, (□ and ■) PEG-2000, (◇ and ◆) MPEG-2000. Open and filled symbols refer to Ce(IV) absorbance and CO₂ evolution rate, respectively.

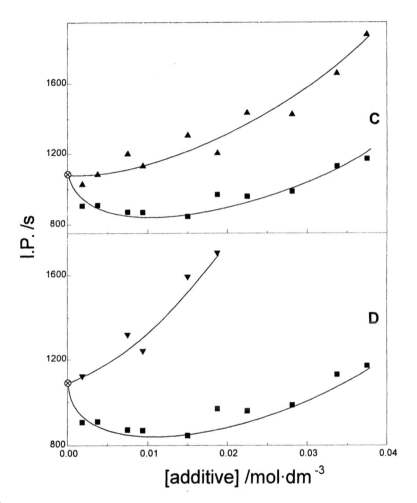

Figure 2. *Continued.*

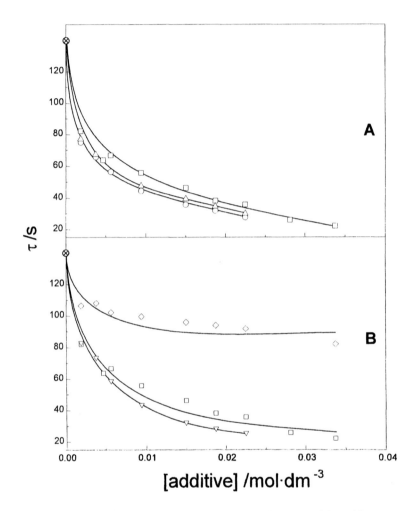

Figure 3. Plot of the oscillation period (τ) as a function of the additive concentration for the perturbed BZ system with (∇ and ▼) EG, (△ and ▲) PEG-400, (○) PEG-900, (□ and ■) PEG-2000, (◇ and ◆) MPEG-2000. Open and filled symbols refer to Ce(IV) absorbance and CO₂ evolution rate, respectively.

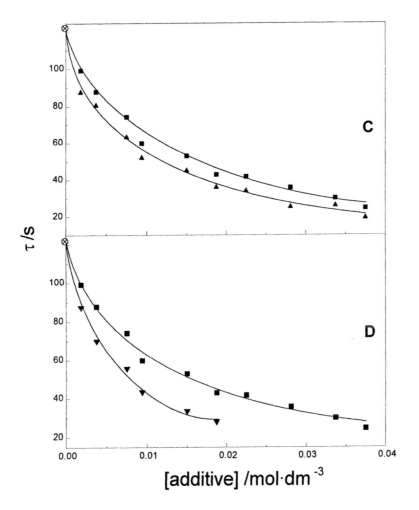

Figure 3. *Continued.*

alcoholic end-groups, but the polymer nature of the perturbant also plays a relevant role.

Thus, when proposing the chemical mechanism, we will separate the perturbation effects into polymer chain (or backbone) and terminal group effects. As we could see the effects caused by polymer backbone reactions are most important at low polymer concentrations but when the PEG concentration is high terminal group reactions dominate.

Reactions of the polymer chain

What type of reactions of the polymer backbone can we expect in a BZ system? First of all it is known (7, 8) that Ce(IV) can react with PEG. The reaction leads to the formation of free radicals

$$Ce(IV) + Poly \longrightarrow Poly\cdot + Ce(III) + H^+ \qquad (P1)$$

The radical formation takes place at the terminal carbon atom linked with the hydroxyl groups and not along the polyether chain. When Ce(IV) is applied in a high excess with respect to the polymer even biradicals can be formed but this was not the case under the experimental conditions used here.

Preliminary kinetic measurements, under the same experimental conditions as used for the BZ, have proved that PEGs react with Ce(IV) on a time scale of few hours depending on the two parameters varied here: the concentration and the molecular weight of the polymer. In particular, the reaction rate increases on increasing either of the above parameters. Moreover, we have also found that the rate for the reaction between MPEG-2000 and Ce(IV) is much lower than that with PEGs and there is no significant change in the Ce(IV) absorbance for days if the reductant is EG. For comparison we also applied dioxane as a reductant, which reacted very slowly. These trends suggest that high molecular weight PEG radicals form faster and more easily than the low molecular weight ones. A systematic kinetic study is in progress.

In analogy to (P1) it is reasonable to assume that beside Ce(IV) the $BrO_2\cdot$ radicals are also able to react with the polymer in reaction (P2):

$$Poly + BrO_2\cdot \longrightarrow Poly\cdot + HBrO_2 \qquad (P2)$$

The two reactions reported above contribute to the production of the radical species Poly· which can scavenge the $BrO_2\cdot$ according to reaction (P3)

$$\text{Poly} \cdot + \text{BrO}_2 \cdot \longrightarrow \text{Poly-BrO}_2 \qquad \text{(P3)}$$

The Poly-BrO$_2$ species so formed can decompose following two different reaction pathways:

$$\text{Poly-BrO}_2 \underset{k_2}{\overset{k_1}{\rightleftarrows}} \begin{array}{l} P_1 + \text{HOBr} \\ \\ P_2 + \text{HBrO}_2 \end{array} \qquad \text{(P4)}$$

This reaction scheme implies that an extra source for the intermediates HOBr and HBrO$_2$ appears in the BZ system. As HBrO$_2$ is the autocatalytic intermediate its production enhances the positive feedback loop, which would lead to an increase in the IP. (This is because "turning off" the autocatalytic HBrO$_2$ production at the end of the IP becomes more difficult.) On the other hand the inflow of HOBr would decrease the IP since HOBr brominates the malonic acid (R31, for the reactions of the unperturbed BZ system we follow the notations of the MBM model):

$$\text{MA (enol)} + \text{HOBr} \longrightarrow \text{BrMA} + \text{H}_2\text{O} \qquad \text{(R31)}$$

and this way increases the bromomalonic acid (BrMA) concentration. BrMA is a part of the negative feedback loop as it is the source of the bromide ions, which inhibit the autocatalytic reaction. To stop the autocatalytic reaction bromide ion and its source bromomalonic acid should reach a critical concentration by the end of the induction period. Consequently, as an inflow of HOBr accelerates the BrMA accumulation during the induction period, the resulting period should decrease accordingly.

The experimental data indicate that the pathway related to k_1 dominates on that occurring *via* k_2 and the extent of the decrease on the IP depends on how easily the given PEG's radicals are formed. In fact, PEG-400 radicals, being formed very slowly, do not contribute to the negative feedback loop, while the other two PEG's radicals significantly do.

Reactions of the terminal –OH groups

The effect of the perturbant at high concentration can be explained by means of the following reactions

$$\text{PEG} + \text{H}^+ + \text{BrO}_3^- \longrightarrow \text{P-CHO} + \text{HBrO}_2 + \text{H}_2\text{O} \qquad \text{(P5)}$$

$$\text{HBrO}_2 + \text{HBrO}_3 \longrightarrow \text{Br}_2\text{O}_4 + \text{H}_2\text{O} \qquad \text{(R6)}$$

$$\text{Br}_2\text{O}_4 \longrightarrow 2\text{BrO}_2\cdot \qquad \text{(R7)}$$

These reactions are analogous to those (*1*) of aliphatic alcohols and EG and justify the overproduction of $\text{BrO}_2\cdot$. Such an *inflow* of the autocatalytic intermediate $\text{BrO}_2\cdot$ will require the system to accumulate a larger amount of BrMA in order to "turn off" the autocatalytic reaction. To reach that higher $[\text{BrMA}]_{\text{CRIT}}$ requires more time, *i.e.* the induction period becomes longer. In the oscillatory regime the inflow of the autocatalytic intermediate will lengthen the autocatalytic and shorten the inhibitory phase. However, as in the present BZ system the inhibitory phase is much longer than the autocatalytic one, resulting in a shortening of the total time period τ.

This explanation is further supported by the experimental evidence that unlike the alcoholic polymer (PEG) MPEG-2000 does not react with acidic bromate and the IP always decreases on increasing MPEG-2000 concentration. Moreover, the experimental trends obtained with PEG perturbed systems for both the oscillation period and the amplitude of oscillations confirm that the reaction with the hydroxyl terminal groups plays a relevant role in perturbing the BZ system. In fact, the presence of MPEG-2000 in the BZ system has a negligible effect on both τ and A in contrast with additives having hydroxyl groups.

Similar trends have been recently (*9*) obtained by Müller et al. for the interaction of poly(vinyl alcohol) (PVA) with the BZ reaction mixture. They compared the effect of PVA with that of isopropanol (IPA), a small molecular weight secondary alcohol whose structure is rather close to that of the polymer repeating unit in PVA. They have found that the secondary alcoholic group of the IPA was a stronger perturbant compared to the same group in PVA even though quantitative modelling of the perturbation effects could not be done mainly for two reasons: the reaction of the BZ subsystem PVA-acidic bromate was found to be too complicated for a systematic kinetic study, and the PVA-IPA comparison was difficult because acetone (which can be brominated) is the product of the reaction between IPA and bromate.

Premixing experiments

In order to compare the rate of reaction (P5) for various perturbants we carried out qualitative experiments where one of the perturbants (at a concentration of $9.37 \cdot 10^{-3}$ mol·dm^{-3}) and all the BZ reagents, except for the Ce(IV), have been premixed for two hours. Our hypothesis was that due to (P5), in the absence of the catalyst, some BrMA should accumulate, which would shorten or eliminate the induction period. It has been found that, for PEG-2000 after two hour premixing, on addition of the cerium catalyst, the system immediately starts to oscillate. For both the other two PEGs used a decrease in the IP was observed, while in the case of MPEG-2000 the IP slightly decreased and with EG the IP is unaffected.

Evidence for a combinative effect of (P2)-(P4) and (P5) was obtained by another premixing experiment. All the BZ reagents (except for Ce(IV)) and MPEG-2000 and EG, both at a concentration of $9.37 \cdot 10^{-3}$ mol·dm^{-3}, were premixed for two hours and on addition of the Ce(IV) the oscillations immediately began. It is reasonable to assume that during the premixing the –OH groups of the EG react with the bromate producing the autocatalytic intermediate BrO$_2$· which now can react further with the polymer backbone of the MPEG. These processes maintain an enhanced bromous and hypobromous acid production during the premixing period. The result is an amplified bromomalonic acid accumulation, which can eliminate the induction period completely. It should be mentioned that premixing of the reactants practically has no influence on the time period and amplitude of the oscillations.

Model calculations

To simulate the response of the oscillatory reaction for polymer perturbations qualitatively we applied i) the latest model of the BZ reaction the Marburg-Budapest-Missoula (MBM) mechanism (4) and added ii) the perturbation reactions of the polymer backbone (P2)-(P4) and iii) of the alcoholic endgroup (P5).

Presently we have no exact values for the rate constants of the polymeric reactions (P2)-(P5); however, these measurements are in progress. Also some modifications in the MBM mechanism can be expected (1,4). Nevertheless, the present semiquantitative calculations applying estimated rate constants prove that the assumed mechanism of the polymeric perturbations qualitatively predicts the observed perturbation effects due to the polymers. Simulation results are shown in Figure 4. In Figure 4A Ce(IV) oscillations of the unperturbed BZ system are depicted. The perturbation effect due to the polymer backbone reaction is shown separately in Figure 4B. It causes a *shortening* of the induction

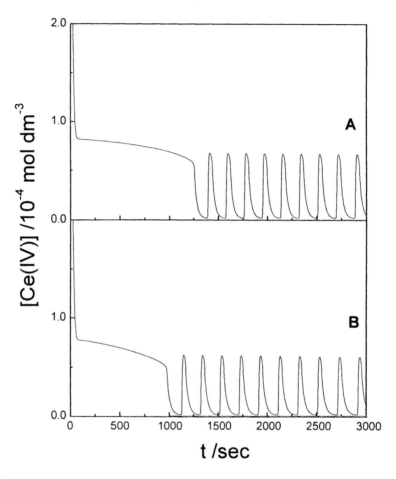

*Figure 4. Ce(IV) concentration vs. time diagrams in a BZ system calculated with
the MBM mechanism. A)Unperturbed system. Initial concentrations [MA] = 0.1
mol·dm⁻³, [KBrO₃] = 0.03 mol·dm⁻³, [Ce(IV)] = 4·10⁻⁴ mol·dm⁻³, [H₂SO₄] = 1
mol·dm⁻³ (See Fig. 1A for comparison). B) The same system as in A) but
perturbed with 0.015 mol dm⁻³ polymer (regarding backbone reactions only).
k_{P2}=400 dm³ mol⁻¹ s⁻¹. It is assumed that the rate determining step of process
(P3)+(P4) leading to HOBr is a diffusion controlled radical-radical
recombination reaction (k_{P3}=10⁹ dm³ mol⁻¹ s⁻¹). C) The same system as in A) but
perturbed with the alcoholic endgroup reaction exclusively. Bromous acid
inflow = 6·10⁻⁸ dm³ mol⁻¹ s⁻¹ due to (P5) was regarded to be independent of time.
D) Combination of the two perturbations shown in B) and C).*

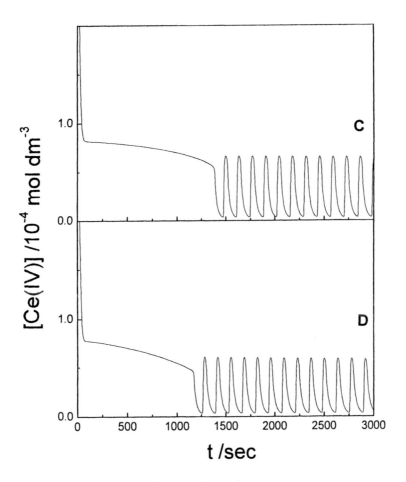

Figure 4. *Continued.*

period in accordance with the experiments and a slight *lengthening* of the time period. A separate perturbation due to the alcoholic endgroups shown in Figure 4C *lengthens* the induction period and *shortens* the time period of the oscillations. Finally a perturbation by both the polymer backbone and alcoholic endgroup reactions causes some shortening of the induction period (this is a result of the two opposite perturbation effects) and a shortening of the time period (again a result of opposite effects but now it is the effect of the alcoholic endgroup, which dominates.)

Acknowledgement

This work was partially supported by OTKA (T-30110 and T-38071) and FKFP-0090/2001 and M.I.U.R (Italy) grants and the ESF Program: "Reactor".

References

1. Pelle, K.; Wittmann, M.; Noszticzius, Z.; Lombardo, R.; Sbriziolo, C.; Turco Liveri, M.L. *J. Phys. Chem. A* **2003**, *107*, 2039.
2. Washington, R.P.; West, W.W.; Misra, G.P.; Pojman, J.A. *J. Am. Chem. Soc.* **1999**, *121*, 7373.
3. Cavasino, F.P.; Sbriziolo, C.; Turco Liveri, M.L.; *Effects of polymers on the Cerium (IV)-catalyzed Belousov-Zhabotinsky reaction*, Communication at the NIMC-EAPS International Conference, Moscow (Russia), 1999. Conference Proceedings (Yamaguchi, T., Ed.) p.10. The abstract focusses on the effect of poly(propylene glycol) (PPG). Experimental details on PEG were given in the full lecture.
4. Hegedüs, L.; Wittmann, M.; Noszticzius, Z.; Shuhua, Y.; Sirimungkala, A.; Försterling, H-D.; Field, R.J. *Faraday Discussions* **2001**, *120*, 21.
5. Försterling, H. D.; Lamberz, H.J.; Schreiber, H., *Z. Naturforsch.* **1983**, *38a*, 483.
6. Nagygyőry, Sz.; Wittmann, M.; Pintér, Sz.; Visegrády, A.; Dancsó, A.; Nguyen Bich, T.; Noszticzius, Z. *J. Phys. Chem. A.* **1999**, *103*, 4885.
7. Harris, J.M.; *Poly(ethylene glycol) Chemistry*; Plenum Press: New York and London, 1992
8. Odian, G.; *Principles of Polymerization*, 3rd edition; John Wiley & Sons, Inc.: New York, 1991, 219.
9. Horváth, J.; Nagy-Ungvárai, Zs.; Müller, S.C. *Phys. Chem. Chem. Phys,* **2001**, *3*, 218.

Chapter 24

Nonlinear Dynamics in Continuous Stirred Tank Reactor Polymerization

Georgia Papavasiliou and Fouad Teymour*

Department of Chemical and Environmental Engineering, Illinois Institute of Technology, Chicago, IL 60616

The Continuous Stirred Tank Reactor (CSTR) has provided a chemical paradigm for nonlinear complex dynamics for almost a century. Advances in this regard are reviewed with special emphasis on polymerization.

Nonlinearity in nature is becoming increasingly evident even to the nonchalant gaze of a lay observer. Inspired by "Fractal Geometry" concepts introduced by Mandelbrot (*1*), we have come to discern order in systems that radically depart from our traditional Euclidian Geometry state of mind. Trees, rivers, lungs, blood vessels and coastlines have incessantly displayed "self-similarity" over the ages, but only now are we capable of detecting and appreciating its beauty and functionality; our perception of spatial order has been altered for good. On the other hand, the popularization of, what some call, chaos "theory" by James Gleick's simplified exposition on the subject (*2*) has changed our perception of temporal order as well. Since the dawn of time, we have lived in a cyclic environment where the annual excursion around the sun forces periodicity in our seasons and consequently in every aspect of our daily lives and even in the rhythms of our own bodies. As humans we have long been fascinated by the apparent cyclic nature of time, but only now are we capable of casting the shroud of formal understanding over this concept and of elucidating the nature of aperiodic phenomena, thus delineating the boundary between the truly random and the deterministically chaotic. Advances in the field of nonlinear dynamics have come from chemistry, physics, biology, economics, sociology and engineering, among others. One modest but important contribution is due to "Chemical Reactor Theory". This short treatise is intended to summarize and analyze the role of a specific class of reactor systems, namely the Continuous Stirred Tank Reactor (CSTR), in the advancement of this field over the past few decades. While this chapter is actually more narrowly focused on a specific polymerization chemistry, the more general impact of the CSTR will be thoroughly reviewed, however the reader should be apprised that this is by no means an exhaustive account of nonlinear phenomena in CSTRs. The works reviewed here are cited as pertains to their contribution to the specific ideas analyzed herein. For a broader review of nonlinear chemical dynamics, the reader is referred to the recent textbook by Epstein and Pojman (*3*).

The Roots of Complexity in the CSTR

The CSTR is an open system exchanging mass and energy with its environment. It is said to achieve a steady state whenever its state variables are invariant in time; this is also often referred to as a fixed-point solution. Under such conditions, the transient balance for each chemical component in the CSTR, $\frac{dC_A}{dt} = \frac{1}{\theta}(C_{Af} - C_A) + r_A$, as written for a reactant A, can be reduced to the steady state equivalent expression $0 = \frac{1}{\theta}(C_{Af} - C_A) + r_A$, where r_A represents the rate of net production of A in the CSTR and θ is the average residence time, while C_A and C_{Af} are the concentrations of A in the reactor and the feed, respectively. For a general *nth* order reaction involving the sole reactant A, the rate of reaction can be expressed in the Arrhenius form as a function of concentration and reactor temperature T:

$$-r_A = k_0 e^{-\frac{E}{RT}} C_A^n \tag{1}$$

where E is the reaction activation energy and k_0 the pre-exponential factor for the reaction rate constant, and where the negative sign indicates the "consumption" of A. The balance equation illustrates the true function of the CSTR at steady state as it indicates that this open system balances the rate of consumption of the reactant A in the reactor against its net rate of supply to the reactor. It is this ability of open systems to sustain a state that is *far from thermodynamic equilibrium* (which in this case would have been the total consumption of A) that gives rise to interesting dynamic phenomena. Note however that complexity does not ensue until this open system feature is coupled with strong non-monotonic nonlinearity. For example, for an isothermal system with a constant temperature T, the rate of reaction given by eq 1 is a monotonically increasing function of C_A over the domain $C_A>0$, while the rate of reactant supply is a linearly decreasing function of the same. As a result, there can only be a unique physically significant solution for the steady state balance. Thus nonlinearity alone is not sufficient for steady state multiplicity; it must be coupled with a degree of non-monotonic feedback on the rate of reaction. The rate of reaction must show a local extremum as a function of extent of reaction. This situation arises when the reaction involved is exothermic, as thermal energy is a product of the reaction. As the extent of reaction increases the reactant concentration decreases but the reactor temperature increases (assuming an adiabatic reactor). Thus the rate of reaction (eq 1) will consist of the product of two terms moving in opposing directions as the extent of reaction increases, leading to a maximum rate at a specific conversion level. Under these conditions, up to three distinct steady state solutions can be obtained. This rate acceleration effect belongs to the general class of autocatalysis, which involves the enhancement of the reaction rate by a product of the reaction. In this specific case the effect is referred to as autothermal behavior, example studies can be found in (4-16). Another class of reactions that exhibit similar behavior is that of chemical autocatalysis. In an autocatalytic reaction, the reaction is catalyzed by a chemical reaction product and its rate increases as the product concentration increases. For example, the cubic autocatalytic reaction: A + 2B → 3B exhibits non-monotonic dependence of the reaction rate on extent of reaction and consequently multiplicity of steady states (17).

Steady state multiplicity is an important nonlinear dynamic phenomenon as it introduces multiple attractors in the phase space, each with a different basin of attraction and different stability characteristics. For each of the cases discussed above, the CSTR was described by at least two active state variables in regions where multiplicity arises; these were C_A, T and C_A, C_B, respectively. Thus the dynamical system possesses two eigenvalues, which introduces the possibility of oscillatory dynamics. In general, a host of oscillatory limit cycle phenomena is encountered in such systems and are well known for the CSTR, however the two variants discussed above can be shown to be devoid of oscillations. In either the Adiabatic CSTR or the CSTR with Cubic Autocatalysis, the dynamic equations can be modified by using an alternative state variable that is a combination of the original two. This is a direct consequence of the fact that the state variables in each variant are directly tied together through the physical process; in the adiabatic CSTR the rate of energy exchange is tied to the rate of mass exchange through convective inflow and outflow, while in autocatalysis A and B are tied through stoichiometry. The resulting equation is linear and leads to the presence of one eigenvalue (often referred to as a washout eigenvalue) at $\lambda = -(1/\theta)$ for all values of parameters. As a result the stability of the CSTR will be solely dictated by the other eigenvalue. The only instability that could result is of the saddle type, which in the phase plane can be seen to define the basin of attraction of the other steady states by its own inset (also known as the separatrix).

Oscillatory phenomena can result in these setups only when additional processes are introduced to break the algebraic bond between the state variables. In each of the cases discussed, this is achievable and has been studied in the literature in detail, as will be seen later. Perhaps, the most important paradigm of complex nonlinear dynamics in chemical reactor theory is the non-adiabatic CSTR. In this setup, an additional means of energy transfer is provided in the reactor (e.g. a cooling jacket or coil), which creates an independent way of decoupling the rates of mass and energy exchange. Both eigenvalues of the system are free to change in response to changes in the parameters and thus complex conjugate pairs can be obtained and oscillatory phenomena will emerge. Stable and unstable limit cycles can thus be encountered and can lead to a host of oscillation varieties in the phase space. The same can be achieved for the autocatalytic reaction in a CSTR if the dynamics of A and B are decoupled by the introduction of a catalyst decay reaction that independently alters the concentration of B in the reactor (17).

By virtue of the Poincaré-Bendixson theorem, no chaotic dynamics can be found in any of the systems discussed so far, because their dynamics are confined to the plane. The emergence of chaos in such systems requires the presence of at least three *active* eigenvalues. Thus a third dimension must be operating in the CSTR at a minimum to introduce the possibility of aperiodic phenomena. Various attempts at observing such phenomena have been conducted including the use of polymerization chemistry, which can include active participation of the dynamics of monomer and initiator concentrations alongside the temperature state variable. Advances in this regard are the focus of this treatise and will be reviewed extensively in the following sections, but the reader should be apprised that other CSTR variants could also lead to chaotic dynamics. Notable among these are the study of Planeaux and Jensen (18), which incorporates the dynamics of the cooling jacket and the work of Kevrekidis et al. on oscillatory periodic forcing of the CSTR (19).

The first observation of multiple steady states in the CSTR was reported by Van Heerden in 1953 (4). He demonstrated that the stability of AIChE J. reactors could be analyzed by simple diagrams of heat production and consumption. However Liljenroth had, unbeknownst to Van Heerden, reported similar phenomena and methodology for catalytic wires in 1918. This work remained unnoticed until 1970 when Luss rediscovered it, as noted by Farr and Aris (5). Later, Bilous and Amundson (6) and Aris and Amundson (7-9) analyzed chemical reactor stability and reported the first possibility of oscillatory behavior in CSTRs. They used Liapunov and Poincaré theories to demonstrate phase plane changes as a function of controller gain for a non-isothermal CSTR with a first order reaction. Uppal Ray and Poore (10, 11) further carried out extensive analysis of reactor dynamics and oscillatory behavior in a CSTR using the methods presented by Poore (12). Multiplicity and stability properties of the CSTR were investigated for a first order reaction as a function of the Damkohler number (10) and the residence time (11). Results demonstrated the existence of five types of steady state bifurcation behavior and seventeen types of dynamic behavior. In later studies, Cohen and Keener (13) focused on multiplicity and stability of oscillatory states in a CSTR with consecutive reactions, Svoronos et al. (*14*) considered the multiplicity of two identical stirred tanks in sequence, and more recent studies by Li (*15*) and Russo and Bequette (*16*) analyzed other variants.

Autocatalytic systems have been studied extensively in Chemical Engineering research for the last two decades (*17, 20-29*) and have been mostly limited to the case of the isothermal CSTR with a single autocatalytic reaction. Gray and Scott (*18*) first analyzed these systems and established that the range of complex phenomena they exhibit is comprehensively similar to those encountered in the autothermal CSTR. As a result of their work and others', this system has become a widely acceptable paradigm for chemically reacting systems that exhibit a rich spectrum of static and dynamic complexity. The absence of stiffness in the model equations of this system has allowed it to catch up to the popularity of the other long standing paradigm, the classical autothermal CSTR. In recent studies, Birol, Teymour and others (*30-33*) have utilized autocatalysis as a paradigm for ecology and population dynamics. They have analyzed the bifurcation behavior of competitive autocatalators that populate a CSTR environment and compete for a common resource (*30-32*), and explored the effect of environment partitioning by considering networks of interactive CSTRs populated by two or more species (*33*).

Polymerization in a CSTR

The dynamic behavior of continuous free-radical polymerization reactors has drawn considerable interest for over four decades. According to Ray and Villa (*34*), 100 million tons of synthetic polymers are produced annually by processes of which a greater part exhibits nonlinear dynamic phenomena. This behavior can occur under both isothermal and non-isothermal conditions in homogeneous (e.g., solution or bulk) and heterogeneous (e.g., emulsion) reactors. This chapter focuses on homogeneous free radical polymerization. For a typical free-radical polymerization kinetic scheme involving initiation, propagation and termination, the rate of polymerization is given by the rate of propagation $r_p = k_p M Y_0$, where M and Y_0 are the monomer and radical concentrations. Using the quasi-steady state assumption, this can be related to the monomer and initiator species, reactor temperature and the rate constants for initiation, propagation and termination, (k_d, k_p and k_t, respectively) and the initiator efficiency, f, as

$$r_p = \underbrace{\left(\frac{k_{p0}k_{d0}^{1/2}f^{1/2}}{k_{t0}^{1/2}}\right)}_{1}\underbrace{\left(e^{\overbrace{\frac{-(E_p + 1/2\,E_d - 1/2\,E_t)}{RT}}^{2}}\right)}_{}\underbrace{[M]}_{3}\underbrace{[I]^{1/2}}_{4} \qquad (2).$$

Table I. Classification of Nonlinear Effect Combinations in eq 2

Active terms in eq 2	Resulting Process	Maximum multiplicity	Chaos Possible?	References
2 & 3	Autothermal	3	No	40, 42, 44-46, 48, 50, 53, 54
1 & 3	Auto-acceleration	3	No	40-43, 45
1, 2 & 3	Autothermal with auto-acceleration	5	No	42, 46, 55
2, 3 & 4	Autothermal with initiator dynamics	3	Yes	51, 57-61

Compared to the rate expression of eq 1, four terms in eq 2, labeled 1 to 4, can combine to affect steady state multiplicity, oscillatory behavior and/or chaotic dynamics. Free-radical polymerization processes are highly exothermic and thus exhibit strong autothermal behavior (terms 2 and 3). However, even under isothermal conditions, sources of non-monotonic nonlinearity arise. One such effect is attributed to the Tromsdorff gel effect whereby viscous reaction conditions cause the rate of polymerization to increase at high conversions, as the radical termination rate becomes diffusion controlled. The termination reaction, which involves the combination of two large radicals, is affected to a greater extent by these limitations than the propagation step involving the diffusion of a small monomer molecule to a large polymer radical. The termination rate constant decreases as monomer conversion increases, thus the product of terms 1 and 3 results in "auto-acceleration" of polymerization rate. Appreciable auto-acceleration can lead to isothermal steady state multiplicity and limit cycle oscillations. Whenever autothermal effects are coupled with significant AIChE J. (the product of terms 1, 2 and 3), two local maxima appear in the rate of reaction and the maximum number of steady states increases to five. All three cases discussed so far are generally confined to two dimensions. If the dimensionality of the system is further increased by the coupling for example of active initiator dynamics along with autothermal effects, (terms 2, 3 & 4) then aperiodic phenomena leading to chaos may emerge as a third eigenvalue participates in the dynamic process. Table I summarizes these combinations of effects along with the literature studies reporting them. Some of the earliest modeling studies of polymerization dynamics, which appeared in the 1960s, cannot be easily classified into these categories either because they involve heterogeneous media or highly idealized kinetic schemes. Of these, the study of Ziegler-Natta ethylene polymerization by Hoftyzer and Zweitering (35) seems to provide the first report of multiple steady states for free-radical continuous polymerization. Their model predicted the presence of five steady states under certain reactor conditions. Warden et al. (36-38) analyzed the stability and control of addition polymerization in a CSTR using Liapunov theroy. Goldstein and Amundson (*39*) analyzed the dynamic and steady state behavior of various free-radical polymerization models in single-phase and two-phase continuous reactors. They demonstrated the existence of up to five steady states for spontaneous initiation of a first order reaction step, and attributed it to strong heat release caused by frequent radical initiation and termination.

Isothermal multiplicity resulting from *Auto-acceleration* was first reported by Jaisinghani and Ray (*40*) for the homopolymerization of methyl methacrylate. They predicted a maximum multiplicity of three steady states associated with the strong gel effect of methyl methacrylate. Schmidt and Ray (*41*) further elaborated this model for solution polymerization and provided the first experimental evidence of multiplicity in homogenous polymerization reactors. Isothermal experiments indicated ignition points and hysteresis phenomena to be in good agreement with model predictions with no adjustable parameters. Model predictions however demonstrated that oscillatory behavior is not expected to occur under isothermal conditions. Adebekun et al. (*42*) also investigated the solution polymerization of methyl methacrylate. Their modeling study concluded that multiplicity attributed to AIChE J. is only present below certain solvent volume fraction. An increase in solvent volume fraction reduces the gel-effect and eliminates multiplicity. They also pointed out that the AIChE J. effect is "directly and nonlinearly" dependent on the isothermal operating temperature. Results showed that reducing temperature resulted in the appearance of a "bended-knee" type of multiplicity.

Jaisinghani and Ray (*40*) also predicted the existence of three steady states for the free-radical polymerization of methyl methacrylate under *autothermal* operation. As their analysis could only locate unstable limit cycles, they concluded that stable oscillations for this system were unlikely. However, they speculated that other monomer-initiator combinations could exhibit more interesting dynamic phenomena. Since at that time there had been no evidence of experimental work for this class of problems, their theoretical analysis provided the foundation for future experimental work aimed at validating the predicted phenomena. Later studies include the investigations of Balaraman et al. (*43*) for the continuous bulk copolymerization of styrene and acrylonitrile, and Kuchanov et al. (*44*) who demonstrated the existence of sustained oscillations for bulk copolymerization under non-isothermal conditions. Hamer, Akramov and Ray (*45*) were first to predict stable limit cycles for non-isothermal solution homopolymerization and copolymerization in a CSTR. Parameter space plots and dynamic simulations were presented for methyl methacrylate and vinyl acetate homopolymerization, as well as for their copolymerization. The copolymerization system exhibited a new bifurcation diagram observed for the first time where three Hopf bifurcations were located, leading to stable and unstable periodic branches over a small parameter range. Schmidt, Clinch and Ray (*46*) provided the first experimental evidence of multiple steady states for non-isothermal solution polymerization. Their experimental findings and model predictions illustrated the existence of isola-type multiplicity for methyl methacrylate and vinyl acetate in solution polymerization. As pointed out by Ray (47), stable steady states on the isola can achieve high conversion at 10% of the reactor size required for the same conversion on the main branch. However, these isolated steady states require special start-up procedures as well as reliable control operation to prevent reactor extinction as inferred by parametric sensitivity analysis. Teymour and Ray (48) reported the first experimental evidence of the existence of limit cycle behavior for this class of reactors for the non-isothermal solution polymerization of vinyl acetate. Figure 1** presents experimental temperature and conversion history of a periodic attractor illustrating large-amplitude sustained oscillations for five reproducible cycles. Results showed that the model accurately predicts the experimental findings of stable steady states as well as the amplitude and period of stable limit cycles collected over a wide range of conditions. A detailed dynamic stability and bifurcation analysis of their model for a lab scale reactor revealed the presence of different bifurcation structures such as s-shape, isola, and mushroom type multiplicities. Among the structures presented, a new bifurcation diagram was observed for the very first time involving a periodic branch with double homoclinic termination. It was concluded that the reactor dynamics are sensitive to changes in operating conditions in the region of parameter space where this structure was located. More extensive experimental evidence of limit cycles over a wider range of reactor conditions was later presented (50). After experimental validation of the model for the lab scale reactor, a model accounting for the difference between full-scale and lab-scale reactors was introduced (51). Even though multiple isolas had been previously reported (42, 52) this was the first account of the coexistence of isolas of periodic solutions with static isolas on the same bifurcation diagram. Russo and Bequette (53) modeled the effect of the dynamics of the cooling jacket on the steady state and dynamic behavior of continuous polymerization reactors.

Copolymerization systems offer an even higher degree of dynamic complexity that involves multiple rates of monomer consumption of the form of eq 2. Pinto and Ray (54) were first to present a comprehensive mathematical model for copolymerization validated with experimental findings. Both experiments and model predictions demonstrated that the stability of such systems is extremely sensitive to small changes in feed composition. For example, when the comonomer feed was switched from pure vinyl acetate to a comonomer mixture containing 0.25 vol% methyl methacrylate, reactor operation changed from stable to oscillatory; rapid convergence to stable operation was observed when the feed composition was switched back.

The coupling of *auto-acceleration* and *autothermal* effects is illustrated in the study of Schmidt et al. (46), which predicted a multiplicity of up to five

---

** All figures are taken from Teymour (49), and are previously unpublished.

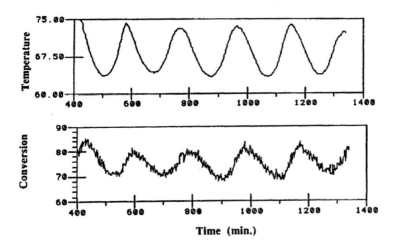

Figure 1. Experimental observation of limit cycle oscillations for vinyl acetate polymerization as reported in (49).

steady states for the non-isothermal system. The additional two steady states were attributed to the strong gel-effect of methyl methacrylate. Interestingly, their five steady states reflected five different conversion values but only three temperature values. This resulted because three of the five steady states occurred at the boiling point of the polymerization mixture. For the same system, Adebekun et al. (42) predicted the existence of a new type of bifurcation behavior involving multiple isolas and the unusual multiplicity structure referred to as the "isola-tuck" which led to a region of five steady states when AIChE J. effects were exacerbated by auto-acceleration. They speculated that the similar distorted s-shape curve that resulted in five steady states reported by Schmidt et al. (46), could have been a predecessor to the formation of an isola in the form of an "isola-tuck." Their theoretical predictions for the non-isothermal polymerization of methyl methacrylate were consistent with both modeling and experimental findings of Schmidt et al. (46), but they extended their study to include the first detailed investigation of the influence of parameters and multiplicity patterns on the molecular weight distribution.

In a bifurcation analysis of a lab-scale copolymerization reactor, Pinto and Ray (55) illustrated interesting bifurcation structures such as the presence of multiple isolas attributed to the methyl methacrylate gel effect. In addition, five-state multiplicity was observed and appeared to be sensitive to the methyl methacrylate comonomer feed concentration. This multiplicity region shrunk with an increase in methyl methacrylate concentration, however, it persisted at very low solvent concentrations. It was also found that more interesting dynamic phenomena occurred as a result of increasing the methyl methacrylate concentration, but these were confined to small regions of parameter space. Furthermore, they predicted that instabilities disappeared for a methyl methacrylate feed concentration exceeding 5 vol%. They pointed out that similar phenomena could be observed for systems where the cross termination rate is higher than that for the respective homo-termination. The five steady state solutions also appeared to be sensitive to changes in the strength of the boiling point constraint, which indicated that the additional steady states were related to both the kinetics and thermodynamics of the reactive system.

Studies of Chaotic Dynamics

As previously mentioned, the dimensionality of the system of equations governs the range of complexity of the expected dynamic behavior. If the dimensionality of the system is embedded in three-dimensional space, then aperiodic phenomena leading to chaos can emerge. Period doubling bifurcations[*] leading to chaotic regimes were found by Teymour and Ray (51),

---

[*] For a primer on chaos terminology and methods cf. Schuster (56) or other textbook.

(*57*). In a detailed analysis of the chaotic regime, they were the first to report chaos in polymerization reactors and were also able to identify a new route to chaos (*51*). They pointed out that relaxing the boiling point constraint on the energy balance resulted in higher reactor temperature and higher levels of initiator conversion, thus making the initiator dynamics more active (*57*). Bifurcation analysis of their full-scale industrial reactor model revealed more complex phenomena such as multiple stable periodic orbits, isolas of periodic solutions and period doubling bifurcations leading to chaos. Figures 2 and 3, taken after Teymour (*49*), illustrate two of these interesting phenomena. Figure 2 presents a return point histogram for the reactor temperature in a chaotic regime that borders a periodic window with a P5 orbit. The points represent the return values encountered on a Poincaré section and show a random alternation between intermittent bursts of chaos and long periods of near periodicity. Figure 3, on the other hand, recreates an iterate map, in a different region of chaotic behavior, by plotting the temperature value of each return to the Poincaré section versus the value at the previous return. This figure illustrates a new route to chaos, akin to the intermittency route, discovered by Teymour and Ray (*51*). At the edges of the periodic windows interspersed in a chaotic regime, the transition from chaotic to periodic behavior usually proceeds through intermittency. Under these conditions the predecessor to a tangent bifurcation appears on the iterate map as a narrow tunnel that traps the dynamics in near periodicity for long periods of time. Teymour and Ray found conditions under which the tangent bifurcation occurs outside the bounds of the strange attractor thus allowing the chaotic attractor to coexist with the periodic attractor in a narrow band preceding the periodic window. The signature of such a transition is evident in Figure 3, which shows that whenever the chaotic attractor is dynamically approached from outside, a long transient of near periodicity is experienced. The circled regions of the map indicate these transients that do not belong to the strange attractor. Kim et al. (*58-60*) also predicted chaotic oscillations for the continuous polymerization of styrene initiated by binary initiator mixtures and by bifunctional initiators. A detailed bifurcation analysis demonstrated that the presence of a second initiator increases the complexity of the dynamics and leads to period doubling bifurcations. They also concluded that below a certain solvent volume fraction no oscillatory behavior was to occur, however, increases above this level resulted in period doubling bifurcations and homoclinic orbits. Chaotic regimes were also investigated by Pinto (*61*) who extended the findings of (*54-55*) to the case of industrial scale copolymerization reactors. The complex dynamic behavior encountered included period doubling bifurcations that led to an accumulation of periodic solutions, chaotic and toroidal oscillations.

It should be mentioned that a considerable amount of research activity is dedicated to multiplicity of steady states and nonlinear dynamics of heterogeneous polymerization systems, especially emulsion polymerization. In

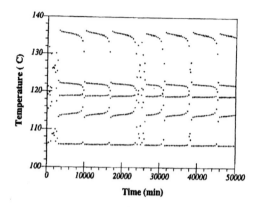

Figure 2. Return point histogram illustrating chaotic intermittency, taken after (49).

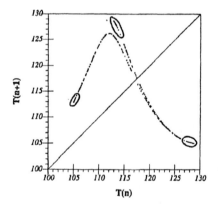

Figure 3. Return iterate map illustrating the hysteresis route to chaos, taken after (49). Circled regions represent transients.

addition to isothermal multiplicity, these multiphase reactors frequently exhibit sustained oscillations also observed in industrial processes due to the particle nucleation and growth phenomena encountered in the polymerization process. Interested readers can find further information in the cited references (*62-73*).

Non-Linear Polymerization

None of the studies mentioned in this review have reported on the dynamic and steady state behavior of continuous *nonlinear* free-radical polymerization processes that could lead to gelation. Papavasiliou (*74*) provides the first instance of an analysis that studies the multiplicity and dynamic behavior of the sol polymer properties and gel fraction for bulk, non-isothermal, free-radical vinyl-divinyl copolymerization in a continuous reactor where crosslinking and gelation are attributed to the propagation of a radical through a pendant double bond. Her analysis uses the "Numerical Fractionation" technique of Teymour and Campbell (*75*). A bifurcation analysis for this polymerization system has revealed s-shape, mushroom, and isola-type multiplicity structures. Gel fraction bifurcation diagrams demonstrating the existence of three steady states have also been found in this study. The analysis illustrates that under the influence of autothermal effects, multiple critical gel transitions are observed. In one instance, gel transitions are located on all three branches of steady states, indicating that gelation can be encountered by either increasing or decreasing the bifurcation parameter, under different conditions. A more detailed dynamic and steady state analysis for this polymerization system will be presented elsewhere.

Conclusions

Attractive in its simplicity, yet complex in its behavior, the Continuous Stirred Tank Reactor has, for the better part of a century, presented the research community with a rich paradigm for nonlinear dynamics and complexity. The root of complex behavior in this system stems from the combination of its open system feature of maintaining a state far from equilibrium and the nonlinear non-monotonic feedback of various variables on the rate of reaction. Its behavior has been studied under various designs, chemistries and configurations and has exhibited almost every known nonlinear dynamics phenomenon. The polymerization chemistry has especially proven fruitful as concerns complex dynamics in a CSTR, as attested to by the numerous studies reviewed in this chapter. All indications are that this simple paradigm will continue to surprise us with many more complex discoveries to come.

References

1. Mandelbrot, B.; *The Fractal Geometry of Nature;* W. H. Freeman: San Francisco, CA, 1983.
2. Gleick, J.; *Chaos: The Making of a New Science;* Viking Press: 1987.
3. Epstein, I.R.; Pojman, J.A. *An Introduction to Nonlinear Chemical Dynamics*; Oxford University Press: New York, NY, 1998.
4. Heerden, C.V. *Ind. Eng. Chem.* **1953**, *45*, 1242-1247.
5. Farr, W.W.; Aris, R. *Chem. Eng. Sci.* **1986**, *41*, 1385-1402.
6. Bilous, O.; Amundson, N.R. *AIChE J.* **1955**, *1*, 513-521.
7. Aris, R.; Amundson, N.R. *Chem. Eng. Sci.* **1958**, *7*, 121-131.
8. Aris, R.; Amundson, N.R. *Chem. Eng. Sci.* **1958**, *7*, 132-147.
9. Aris, R.; Amundson, N.R. *Chem. Eng. Sci.* **1958**, *7*, 148-155.
10. Uppal, A.; Ray, W.H.; Poore, A.B. *Chem. Eng. Sci.* **1974**, *29*, 967-985.
11. Uppal, A.; Ray, W.H.; Poore, A.B. *Chem. Eng. Sci.* **1976**, *31*, 205-214.
12. Poore, A.B. *Arch. Ratl. Mech. Anal.* **1973**, *52*, 358-388.
13. Cohen. D.; Keener, J.P. *Chem. Eng. Sci.* **1976**, *31*, 115-122.
14. Svoronos, S.; Aris, R.; Stephanopoulos, G. *Chem. Eng. Sci.* **1982**, *37*, 357-366.
15. Li, R. *Chem. Eng. Sci.* **1994**, *18*, 3053-3057.
16. Russo, L.P.; Bequette, B.W. *AIChE J.* **1995**, *41*, 135-147.
17. Gray, P.; Scott, S.K. *Chem. Eng. Sci.* **1983**, *38*, 29-43.
18. Planeaux, J.B.; Jensen, K.F. *Chem. Eng. Sci.* **1986**, *41*, 1497-1523.
19. Kevrekidis, I.G.; Aris, R.; Schmidt, L.D. *Chem. Eng. Sci.* **1986**, *41*, 1549-1560.
20. Scott, S. K. *Chem. Eng. Sci.* **1983**, *38*, 1701-1708.
21. Gray, P.; Scott, S.K. *Chem. Eng. Sci.* **1984**, *39*, 1087-1097.
22. Balakotaiah, V. *Proc. Royal Soc., Ser. A: Math. Phys. Sci.* **1987**, *411*, 193-206.
23. Kay, S.R.; Scott, S.K.; Tomlin, A.S. *Chem. Eng. Sci.* **1989**, *44*, 1129-1137.
24. Adesina, A.A.; Adewale, K.E.P. *Ind. Eng. Chem. Res. 1991*, *30*, 430-434.
25. Liou, C.-T.; Chien, Y.-S. *Chem. Eng. Sci.* **1995**, *50*, 3637-3644.
26. Chien, Y.-S.; Liou, C.-T. *Chem. Eng. Sci.* **1995**, *50*, 3645-3650.
27. Kumar, J.; Nath, S. *Chem. Eng. Sci.* **1997**, *52*, 3455-3462.
28. Adesina, A.A. *Dev. Chem. Eng. Mineral Proc.* **1998**, *6*, 135-152.
29. Parulekar, S. J. *Chem. Eng. Sci.* **1998** 53, 2379–2394.
30. Birol, I.; Teymour; F. *Physica D* **2000** 144, 279–297.
31. Chaivorapoj, W.; Birol, I.; Teymour, F. *Ind. Eng. Chem. Res.* **2002** *41*, 3630-3641.
32. Chaivorapoj, W.; Birol, I.; Teymour, F.; Çinar, A. *Ind. Eng. Chem. Res.* **2003**, in press.
33. Birol, I.; Parulekar, S.J.; Teymour, F. *Phys. Rev. E* **2002** *66*, 051916.
34. Ray, W.H.; Villa, C.M. *Chem. Eng. Sci.* **2000** 55, 275-290.
35. Hoftyzer, P.J.; Zwietering, T.N. *Chem. Eng. Sci.* **1961**, *12*, 241-251.
36. Warden, R.B.; Amundson, N.R. *Chem. Eng. Sci.* **1962**, *17*, 725-734.
37. Warden, R.B.; Aris, R.; Amundson, N.R. *Chem. Eng. Sci.* **1964**, *19*, 149-172.
38. Warden, R.B.; Aris, R.; Amundson, N.R. *Chem. Eng. Sci.* **1964**, *19*, 173-190.

39. Goldstein, R.P.; Amundson, N.R. *Chem. Eng. Sci.* **1965**, *20*, 195-236.
40. Jaisinghani, R.; Ray, W.H. *Chem. Eng. Sci.* **1977** *32*, 811-825.
41. Schmidt, A.D.; Ray, W.H. *Chem. Eng. Sci.* **1981** *36*, 1401-1410.
42. Adebekun, A.K.; Kwalik, K.M.; Schork, F.J. *Chem. Eng. Sci.* **1989** *44*, 2269-2281.
43. Balaraman, K.S.; Kulkarni, B.D.; Mashelkar, R.A. *Chem. Eng. Comm.* **1982** *16*, 349-360.
44. Kuchanov, S.J.; Efremov, V.A.; Slin'ko, M.G. *Proc Acad. Soc. USSR: Phys. Chem.* **1986**, *283*, 686-690.
45. Hamer, J.W.; Akramov, T.A.; Ray, W.H. *Chem. Eng. Sci.* **1981**, *36*, 1897-1914.
46. Schmidt, A.D.; Clinch, A.B.; Ray, W.H. *Chem. Eng. Sci.* **1984**, *39*, 419-432.
47. Ray, W.H; *Chemical Reaction Engineering-Plenary Lectures*; Wei, J.; Georgakis, C.; ACS Symposium Series 226; American Chemical Society: Washington, D.C., 1983; Vol. 5, 101-133.
48. Teymour, F.; Ray, W.H. *Chem. Eng. Sci.* **1989** *44*, 1967-1982.
49. Teymour, F.A. Ph.D. Thesis, University of Wisconsin, Madison, WI, 1989.
50. Teymour, F.; Ray, W.H. *Chem. Eng. Sci.* **1992** *47*, 4121-4132.
51. Teymour, F.; Ray, W.H. *Chaos, Solitons and Fractals* **1991** *1*, 295-315.
52. Adomaitis, A.; Cinar, A. *Proc. Am. Control Conf.* **1987**, 1419-1424.
53. Russo, L.P.; Bequette, B.W. *Chem. Eng. Sci.* **1998**, *53*, 27-45.
54. Pinto, J.C.; Ray, W.H. *Chem. Eng. Sci.* **1995**, *50*, 715-736.
55. Pinto, J.C.; Ray, W.H. *Chem. Eng. Sci.* **1995**, *50*, 1041-1056.
56. Schuster, H.G.; *Deterministic Chaos*; Physic-Verlag: Weinheim, 1984.
57. Teymour, F.; Ray, W.H. *Chem. Eng. Sci.* **1992** *47*, 4133-4140.
58. Kim, K.J.; Choi, K.Y.; Alexander, J.C. *Polym. Eng. Sci.* **1990**, *30*, 279-290.
59. Kim, K.J.; Choi, K.Y.; Alexander, J.C. *Polym. Eng. Sci.* **1991**, *31*, 333-352.
60. Kim, K.J; Choi, K.Y. *Polym. Eng. Sci.* **1992**, *32*, 494-505.
61. Pinto, J.C. *Chem. Eng. Sci.* **1995**, *50*, 3455-3475.
62. Gershberg, D.B.; Longfield, J.E. *Kinetics of the Continuous Emulsion Polymerization* **1961**, 54th AIChE Meeting, New York.
63. Omi, S.; Ueda, T.; Kubota, H. *J. Chem. Engng Japan* **1969**, *2*, 193-198.
64. Gerrens, H.; Kuchner; Ley, G. *Chemie-Ing.-Tech.* **1971**, *43*, 693-698.
65. Ley, G.; Gerrens, H. *Makromol. Chem.* **1974**, *175*, 563-581.
66. Kirillov, V.A.; Ray, W.H. *Chem. Eng. Sci.* **1978**, *33*, 1499-1506.
67. Kiparissides, C.; MacGregor, J.F.; Hamielec, A.E. *J. Appl. Polym. Sci.* **1979**, *23*, 401-418.
68. Rawlings, J.B.; Ray, W.H. *Chem. Eng. Sci.* **1987**, *11*, 2767-2777.
69. Rawlings, J.B.; Ray, W.H. *AIChE Journal* **1987**, *33*, 1663-1676.
70. Schork, F.J.; Ray, W.H. *J. Appl. Polym. Sci.* **1987**, *34*, 1259-1276.
71. Penlidis, A.; MacGregor, J.F.; Hamielec, A.E. *Chem. Eng. Sci.* **1989**, *44*, 273-281.
72. Lu, Y.J.; Brooks, B.W. *Chem. Eng. Sci.* **1989**, *44*, 857-871.
73. Araújo, P.H.; Abad, C.; de la Cal, J.C.; Pinto, J.C.; Asua, J.M. *Polym. React. Eng.* **2001**, *9*, 1-17.
74. Papavasiliou, G. Ph.D. Thesis, Illinois Institute of Technology, Chicago, IL, 2003.
75. Teymour F.; Campbell, J. D. *Macromolecules*, **1994**, 27, 2460-2469.

Indexes

Author Index

Subject Index